A Lesson for the Future of Our Science
My Testimony on
Lord Patrick M S Blackett

A Lesson for the Future of Our Science
My Testimony on Lord Patrick M S Blackett

Antonino Zichichi

European Physical Society, Switzerland

NEW JERSEY · LONDON · SINGAPORE · BEIJING · SHANGHAI · HONG KONG · TAIPEI · CHENNAI · TOKYO

Published by

World Scientific Publishing Co. Pte. Ltd.
5 Toh Tuck Link, Singapore 596224
USA office: 27 Warren Street, Suite 401-402, Hackensack, NJ 07601
UK office: 57 Shelton Street, Covent Garden, London WC2H 9HE

Library of Congress Cataloging-in-Publication Data
Names: Zichichi, Antonino, author.
Title: A lesson for the future of our science : my testimony on Lord Patrick M.S. Blackett /
   Antonino Zichichi, European Physical Society, Geneva, Switzerland.
Description: Singapore ; Hackensack, NJ : World Scientific, [2016] | 2016 |
   Includes bibliographical references.
Identifiers: LCCN 2015047282| ISBN 9789814719674 (hardcover ; alk. paper) |
   ISBN 9814719676 (hardcover ; alk. paper) | ISBN 9789814719414 (pbk. ; alk. paper) |
   ISBN 9814719412 (pbk. ; alk. paper)
Subjects: LCSH: Blackett, P. M. S. (Patrick Maynard Stuart), Baron Blackett, 1897–1974. |
   Physicists--Great Britain--Biography. | Creative ability in science. |
   Particles (Nuclear physics) | Vacuum polarization. | World War, 1939–1945--Science.
Classification: LCC QC16.B59 Z53 2016 | DDC 530.092--dc23
LC record available at http://lccn.loc.gov/2015047282

British Library Cataloguing-in-Publication Data
A catalogue record for this book is available from the British Library.

Copyright © 2016 by World Scientific Publishing Co. Pte. Ltd.

All rights reserved. This book, or parts thereof, may not be reproduced in any form or by any means, electronic or mechanical, including photocopying, recording or any information storage and retrieval system now known or to be invented, without written permission from the publisher.

For photocopying of material in this volume, please pay a copying fee through the Copyright Clearance Center, Inc., 222 Rosewood Drive, Danvers, MA 01923, USA. In this case permission to photocopy is not required from the publisher.

In-house Editor: Rhaimie Wahap

Antonino Zichichi, *Emeritus Professor of Advanced Physics at the University of Bologna, has authored over 1100 scientific papers which include: 7 discoveries, 5 inventions, 3 original ideas which opened new avenues in high energy subnuclear physics and 5 high-precision measurements of fundamental physics properties. Among the discoveries are the Effective Energy in QCD and the nuclear antimatter; among the inventions are the electronic circuit for time-of-flight measurements with a precision of seventy picoseconds (thousandths of nanosecond); among the original ideas is that which brought the discovery of the third column in the fundamental structure of the Universe.*

The great projects of European Physics – LEP and LHC at CERN, GRAN SASSO at INFN, HERA at DESY – are all linked to his name for his seminal contributions in their conception consequent study and implementation phases.

He has been in charge at the European and National level (EPS and INFN). He founded the "Ettore Majorana Center for Scientific Culture" in Erice and the "Enrico Fermi Centre" in Rome. He is President of the "World Federation of Scientists".

Nine books were written by eminent scientists about his discoveries and inventions. The asteroid discovered in 1986 has been dedicated to him, 3951 Zichichi. He has written 21 books; received 104 Prizes, 24 honorary citizenships, 10 Gold Medals, 9 honorary Ph.D. degrees and is member of 13 scientific Academies. He was awarded honours in 16 Countries: Argentina, China, Georgia, Germany, Italy, Kyrgyzstan, Lithuania, Malta, Moldova, Poland, Romania, Russia, Ukraine, UK, USA and Vatican City.

Michael Duff
Imperial College, London, UK

WHY OUR SCIENCE

There are many intellectual activities which people consider to be Science. But there is a general agreement about Physics being the mother and the queen of all Sciences.

We should not forget that Modern Science, thanks to Galileo Galilei, is based on the following fundamental point: no matter what we are thinking (including the very rigorous theoretical form of thinking, which means expressing all our ideas using mathematical language, i.e. formulae), we have to find out what experiment has to be realized in order to prove the validity of our intellectual activity. If this is not the case, our intellectual activity has nothing to do with Science.

Richard Feynman (Erice, 1964): "It doesn't matter how beautiful your theory is, it doesn't matter how smart you are. If it doesn't agree with experiment, it's wrong."

WHY MY TESTIMONY

When, thanks to Paul Dirac, I had the privilege of meeting with Pyotr L. Kapitza, he told me that, when he had the courage of refusing the offer by Stalin to be the Director of the USSR nuclear fusion project, losing all material privileges including his salary, he realized that the only treasury nobody could take away from him was what all he knew and kept in his memory.

It happens that Blackett was firmly convinced that we physicists should not waste our time in writing notes in agendas but keeping all relevant steps in our brain, thanks to the privilege of having memory. The roots of all our activities are in our memory.

The present volume would have remained in my memory if it was not for the invitation I received from Mike Duff to give a lecture at Imperial College(*).

*This is how I realized that sometimes we have to **stop** working on what we are still trying to understand better.*

*This **stop** has to be given top priority, at least once in our life, when we have to pay the due tribute to those who have played a crucial role in our scientific and cultural life.*

The content of this book is based on what I can remember of the time when I started my activity in physics, including all consequences in terms of Physics, Logic and Ethics, in our Science. The published papers are the written testimony of what I tried to reconstruct.

(*) "My Testimony on Lord Patrick M.S. Blackett" delivered at the Clore Lecture Theatre – Imperial College, **London, 30th April 2014.**

Lord Patrick M.S. Blackett.

(Picture courtesy of the Imperial College London).

A LESSON FOR THE FUTURE OF OUR SCIENCE
MY TESTIMONY ON LORD PATRICK M.S. BLACKETT

ANTONINO ZICHICHI

University of Bologna and INFN, Italy
CERN, Geneva, Switzerland
Enrico Fermi Centre, Rome, Italy
Pontifical Academy of Sciences, Vatican City
World Federation of Scientists, Beijing, Geneva, Moscow, New York
Ettore Majorana Foundation and Centre for Scientific Culture, Erice, Italy

CONTENTS

I	– *The reasons why this book has been written*	5
II	– *An incredible sequence of Unexpected Events: the Blackett, Bohr and Rabi "vital condition" for CERN*	13
II-1	– *Europe had only cosmic rays*	13
II-2	– *How it could happen that – few years after CERN was established – the discovery needed for the existence of the "strange" charge was obtained*	19
II-3	– *An example of ethics in our Science*	23
III	– *Blackett and the discovery of the Subnuclear Universe*	27
III-1	– *The V–particles: from strangeness to QCD and its "hidden" side plus the "end of a myth"*	27
III-2	– *The Nuclear Glue: from the π–meson to the third Family plus the gluon jets and the instantons*	41
III-3	– *Virtual Physics: from the "vacuum polarization" to the grand unification of all forces and the GAP*	47
III-3.1	– *The 1st discovery at CERN in the Physics of high precision. Consequences of Virtual Physics on the "Heavy Electron" called muon*	61
III-3.2	– *The 1st invention at CERN for the production of high precision magnetic fields that is 10^2 times faster for construction and 10^2 times cheaper than all existing technologies*	65
III-3.3	– *The 1st proof that CERN could compete with well established famous Labs and win*	69
III-3.4	– *The third lepton and the problem of Gödel in Physics*	73
IV	– *Blackett and Russell (Galilei, Einstein, Gödel)*	79
V	– *The "Blackett Effect" in the 2nd World War*	87
VI	– *Blackett, the Cambridge Circle and the whole of our knowledge including Virtual History and the three Big Bangs*	93
VII	– *New Institutions founded*	109
VIII	– *Memory is needed for the Future*	125
IX	– *The Future*	141
X	– *Conclusions – From Blackett to present day Physics*	149
XI	– *The "Piersanti Mattarella Tower of Thought" and the view which enchanted Professor Blackett*	159
XII	– *How things really happen*	171
	Appendices	187
	References	239
	Acronyms	251
	Index of Names	254
	Analytic Index of the Main Topics	261

The central door entrance of the Blackett Institute in Erice.

I – THE REASONS WHY THIS BOOK HAS BEEN WRITTEN

The contributions of Professor Blackett to progress in Physics, in Scientific Culture and in the 2nd World War deserve to be highlighted.

In 1932, it was Blackett who provided the experimental proof for the existence of the so called "vacuum polarization" effect, the first example of "virtual physics". Without the existence of "virtual phenomena" it would have been impossible to have the theoretical structure to develop the Unification of the Fundamental Forces of Nature, called Grand Unified Theory (GUT).

It is the discovery of the so-called "strange particles" in the Blackett group that opened a new horizon towards the existence of the subnuclear universe.

During the 2nd World War, thanks to the "Blackett effect", the British Navy won the Mediterranean battle which gave Sicily two years of peace instead of the two terrible last years of war, from mid-1943 to mid-1945. The Blackett effect will be the subject of Chapter V.

On the problem of developing new institutions after the last attempt of Europe to destroy itself during the 2nd World War, there are two inspiring examples: CERN in Geneva, now the largest and most powerful laboratory investigating the subnuclear universe, and the EMFCSC (Ettore Majorana Foundation and Centre for Scientific Culture) in Erice, now the famous Centre where scientists from all countries work together for a science without secrets and without borders.

CERN is well known all over the world. I will later describe the first activities of CERN in more detail: two discoveries and one invention. **The first Physics discovery**, just a few years after CERN's foundation (in 1954), was the simultaneous production of heavy mesons with positive and negative strangeness (as we will see in Chapter II). **The second Physics discovery** was in the field of high precision experiments on the muon magnetic moment (Chapter III and Section III-3.1). **The first invention** was a new technology for the construction of high precision magnetic fields a hundred times cheaper and a hundred times faster to construct than all existing technologies (Chapter III-3.2). There is also proof that this new laboratory was able to do what other labs tried to do and failed. This was in the measurement of the universal weak charge (Chapter III-3.3), now known as the Fermi coupling. These achievements allowed CERN to be known in the Physics community the world over.

These results could be achieved at CERN because Blackett gave to CERN his cosmic ray group, which was the most powerful experimental Physics group in the world, as we will see in Chapter II.

Concerning the EMFCSC, I will only give some data. During the five past decades, more than 100,000 scientists from all over the world have participated in the 126 International Schools of the EMFCSC. The majority of these scientists were engaged in activities far away from Physics. The data are in the table below.

DATA ON ACTIVITIES OF THE ETTORE MAJORANA FOUNDATION AND CENTRE FOR SCIENTIFIC CULTURE SINCE 1963
126 SCHOOLS,
1,778 COURSES,
123,716 PARTICIPANTS[*]
COMING FROM **932** UNIVERSITIES
AND LABORATORIES OF **140** NATIONS.

At the EMFCSC, the first institute was dedicated to Lord Patrick M.S. Blackett. Many fellows asked me why it was named after Blackett.

---

[*] One hundred and thirty five Nobel Laureates (eighty-six of them were awarded the Nobel Prize after their participation to the EMFCSC Schools and forty-nine were already laureates when they started taking part in the EMFCSC's activities).

Having told them the reason, they encouraged me to write about these "extremely interesting" facts. I have not done it till now. What follows is my first attempt.

Professor Mike Duff organized at Imperial College on April 2014 a formidable set of events and invited me to recall what happened when I was 67 years younger than now (2014).

In fact, the first time I learned about the existence of the Blackett group was in 1947 when the new particles called Vs were discovered by a group of physicists whose leader was the same fellow who gave Sicily the privilege previously mentioned during the 2nd World War of two extra years free from war. 1947 was the last year of my pre-university studies, in Trapani (Ximenes Liceum), Sicily.

I could not imagine that eight years after this totally unexpected discovery I would have become the youngest member of Professor Blackett's Physics group, whose two discoveries – the production of (e^+e^-) pairs in 1932 in London and in 1947 of the V–particles in Manchester – attracted the attention of the physics community the world over. When I joined the Blackett group the detectors (Wilson-cloud-chamber and associated equipment) were installed in the Sphinx Observatory, Europe's highest lab (3,580 meters a.s.l.), at Jungfraujoch (see **Photos 1** and **2**).

Photo 1: The Jungfraujoch complex where the highest Railway Station of Europe, the Research Center and the Sphinx Observatory are located.

Photo 2: The Jungfraujoch Lab (in the Sphinx Observatory) where the Blackett group had its powerful detectors.

From Sicily I went to the top-Lab of Europe, thanks to an incredible sequence of unexpected events which allowed me to become a pupil of Professor Blackett (see Chapter II). **Professor Blackett was the leader of the most powerful group of experimental Physics in the world**. This has been of vital importance not only for my career in Physics but also for my activities, including the scientific culture endeavours, to communicate to people the role of scientific discoveries in everyday life and the establishment of new institutions such as the ones already mentioned – **CERN** in Geneva and **EMFCSC** in Erice – and later the **WFS** (the World Federation of Scientists) in Geneva, New York, Beijing and Moscow and the **World Lab**, with its projects implemented in many developing countries (see Chapter VII).

Professor Blackett was convinced that it is us, the physicists, who must be engaged directly in letting the people outside our labs know what the role of science is in the progress of our civilisation.

I would like to thank Lord Blackett: it is because of him that I had the privilege of spending an evening with his friend **Bertrand Russell** and getting to know his views on us physicists who were and are engaged at the frontiers of human knowledge in order to understand the Logic of Nature (see Chapter IV).

Professor Blackett and his friend Bertrand Russell were interested in identifying the real motor for progress in technological inventions that has allowed the quality of life to be at the level it is today. We will see in Chapter VI that this motor is the scientific discovery at the **1st level of Galilean Science**.

We should all **pay tribute** to a physicist who played a vital role in the discovery of the subnuclear universe and in the promotion of scientific culture.

Today, the centre of everyone's attention are the problems concerning the **role of science** as the source of high quality life instead of being just the source of high precision and high power weapons. In the years after the 2nd World War, these problems were far from being of interest to the public. When I had the privilege of being in Professor Blackett's Physics group I learned a lot about these problems, essentially as Professor Blackett was engaged not only in the frontiers of Physics but also with problems concerning the role of science in the culture of our time (Chapter VI). This is how I learnt about the great technological achievements obtained thanks to the existence of the Manhattan Project, which developed the so-called atomic bomb in the USA during 2nd World War. In a few years the discovery of the "nuclear fission" produced a nuclear fire, millions of times more powerful than all known fires. Why were the nuclear bombs that destroyed Hiroshima and Nagasaki called "Atomic"? The answer was given to me by Professor Wigner, father of Time-reversal invariance, as we will see later. The fellows responsible for the Manhattan Project were three great physicists (Fermi, Oppenheimer, Wigner) and a military exponent of the USA army. When the nuclear fission bomb was experimentally checked to be working as expected, the three physicists proposed to give the name nuclear to the bomb. But the military exponent of the Manhattan Project pointed out that before 1940 the word **"Nuclear" did not exist in our language**. The name given to this invention of enormous power had to be the most advanced scientific word, not an unknown word. People would have been confused. The most advanced word was "atomic" and that was the reason why the first nuclear bomb was called "Atomic" (see *"Mach died convinced that Atomic Physics was not Science"*, Appendix 1).

Now few words on the Time-reversal invariance. For us the time goes from past to future, apparently because we are very complex systems. The simplest examples of particles, called elementary, have no memory and no clocks; their interactions are the results of fundamental forces (see *"Fundamental Forces"*, Appendix 2). For these particles, **whatever they do** must be invariant with respect to time going from the past to the future and from the future to the past. The validity of the Wigner theorem was proved for the electromagnetic forces by A.Z. in the middle of last century, as we will see in Chapter IV.

The Manhattan Project was the result of the collaboration of many brilliant physicists working together in one place. According to Blackett and Russell, this project was the example of how the new frontiers of science and technology would have to be implemented in the future. The Manhattan Project is the proof that a new bridge is not only possible but also needed in order to fill the gap between traditional university teaching and the big projects for the future of science and technology. **CERN** did not exist at that point, nor other institutions such as the **EMFCSC** in Erice and the International Centre for Theoretical Physics (**ICTP**) in Trieste [1]. Abdus Salam (**Photo 3**), the founder of ICTP, had yet to join the Blackett group.

Photo 3: Professor Abdus Salam in Erice (1987).

Blackett and Russell discussed the following point. The first university was founded more than 900 (927 to be precise) years ago in one of Europe's most developed areas, Bologna, in order to avoid waiting ten years before books were available. Medicine, astronomy and law were topics being investigated by few specialists.

Instead of waiting for the specialists to write a book (printing had yet to be invented), why not invite them to give lectures? Now we have instant books but the primary role of the universities has changed. **Why**? Because human knowledge has exploded during the last four hundred years, since the time when Galilei discovered the first Fundamental Laws of Nature [2]. And this produced an exponential growth in human knowledge. This exponential growth is the answer to our **why**. Nowadays, the task of the universities is mainly pedagogical and therefore there is a gap between universities and research laboratories which needs to be filled.

During many years this gap has been filled only for military technologies. In fact, starting with the Uranium (Hiroshima) and Plutonium (Nagasaki) bombs, both based on nuclear-fission, nuclear-fusion bombs (called H-bombs) were realized in 1952 by the USA and in 1953 by USSR, the sequence being synthetized below:

$$
\begin{aligned}
\text{Nuclear Fission} &\rightarrow \begin{cases} 1945 - \text{Uranium Bomb (Hiroshima);} \\ 1945 - \text{Plutonium Bomb (Nagasaki);} \end{cases} \\
\text{Nuclear Fusion} &\rightarrow \begin{cases} 1952 - \text{H-Bomb (USA);} \\ 1953 - \text{H-Bomb (USSR).} \end{cases}
\end{aligned}
$$

This sequence has as its basis the Manhattan Project. No institutions devoted to non-military technologies at that point existed. New institutions entirely devoted to non-military technology were needed – one for the industrialized countries, the **EMFCSC,** and the other for the developing countries, the **ICTP** [3–4]. Both institutions had to fulfil the role of the first university, established nearly a millennium earlier. These initiatives had Blackett as their strongest supporter. This was the origin of the two now well-known centres for advanced studies in Erice (**EMFCSC**) and in Trieste (**ICTP**).

And now, back to the year 1947 when no physicist would have been able to predict the existence of the subnuclear world. Here is the proof. When the V–particles were discovered by the Blackett group, Enrico Fermi said:

"These V–particles are the price

we have to pay for having discovered

probably *everything needed to understand*

the Logic of Nature."

Fermi was an extremely careful scientist. **"Probably"** in his quote above became the incredible series of new, unpredicted and totally unexpected, discoveries implemented during the following decades.

In 1947, with all particles known to exist (p, n, e, π, γ, ν)$^{(*)}$, it was possible to understand the molecular, atomic and nuclear structure of all types of matter: stones, mountains, oceans, the Earth, the Sun, the Moon, the Galaxies. The fundamental forces needed were only four: gravitational, electromagnetic, weak and nuclear (see *"Fundamental Forces"*, Appendix 2). The general trend was: there is not much left still to be discovered. But as I will describe in this book, and thanks to Professor Blackett, the second half of the twentieth century saw that there was so much more to find out about the physical world. The *"so much more"* has its roots in the V–particles.

And now comes the incredible sequence of unexpected events that led to me becoming the pupil of **Professor Blackett**.

(*) Few words for each symbol: p ≡ proton; n ≡ neutron; e ≡ electron; π ≡ nuclear glue called pi-meson; γ ≡ photon, gamma ray; ν ≡ neutrino.
The existence of the neutrino, ν, was predicted much before 1947 by Pauli in order to avoid the energy conservation law to be violated in radioactive decay processes. Its experimental discovery was announced at the Geneva Conference in 1956.

II – AN INCREDIBLE SEQUENCE OF UNEXPECTED EVENTS: THE BLACKETT, BOHR AND RABI *"VITAL CONDITION"* FOR CERN

II-1 – EUROPE HAD ONLY COSMIC RAYS

The year was 1955. Having obtained a fellowship of the INFN[*] I was at the "Guglielmo Marconi" Physics Institute of the Rome University La Sapienza, going up and down from the Cervinia Lab (3,480 meters a.s.l.), since cosmic rays were the only source of High Energy Particles in Europe.

> **Europe** had only **Cosmic Rays**.
> **No** SC
> **No** PS
> **No** ISR
> **No** SPS
> **No** LEP
> **No** LHC.
> see Appendix 3

It was midnight on a Sunday and I was analysing Wilson-cloud-chamber-pictures, taken by our INFN-Rome group in the Cervinia Lab, hoping to discover something new, produced by cosmic rays. There was a phone call from

---

[*] INFN = Istituto Nazionale di Fisica Nucleare: the largest Italian Organization in the Field of Nuclear and Subnuclear Physics.

Bruno Brunelli, the Deputy Group Leader, asking if I could leave the next day to participate at an international conference since he was sick (down with 40° fever).

These were times when each country had a quota for participants. Minus one implied that, at the next conference, the country would have had its quota reduced by one. This was the reason for the "last minute call".

I did go to the conference where there was a big announcement: **a new discovery** by the same group who in 1947 discovered the V–particles.

This was the first name originally given by the group leader, Professor Blackett, to this unexpected discovery. Why V–particles? Because they appeared in the Wilson-cloud-chamber-pictures as inverted V.

The discovery of the Vs was the starting point of the physics of "strange particles", whose consequences no one in 1947 would have imagined. It was to be the opening of a new horizon towards the subnuclear universe.

The first quantity needed was a new quantum number which was called "strangeness" whose symbol was (and is) "s".

The value of this quantum number had to be

$$+1 \text{ for particles}$$

and

$$-1 \text{ for antiparticles.}$$

In all reactions of the subnuclear universe this new quantum number had to be additively conserved. We will see that this was a big problem which needed ten years (1947–1957) to be experimentally solved [5].

The new discovery presented by a member of the Blackett group (G.D. James) needed a reaction where a heavy meson, called θ^0, interacting

II – An Incredible Sequence of Unexpected Events: the Blackett, Bohr and Rabi *"Vital Condition"* for CERN

with a neutron, n, had to produce a baryon Λ^0 plus a pi-meson, π. The reaction needed was therefore

$$\theta^0 + n \rightarrow \Lambda^0 + \pi.$$

But this reaction was "forbidden", since it did not conserve the so-called "strangeness" quantum number.

In **Figure 1** there is a copy of the page in the proceedings where this was reported [G.D. James, *"Some Notes on the Production of V-particles"*, Suppl. Nuovo Cimento Vol. IV, X n. 2, page 325 (1956)].

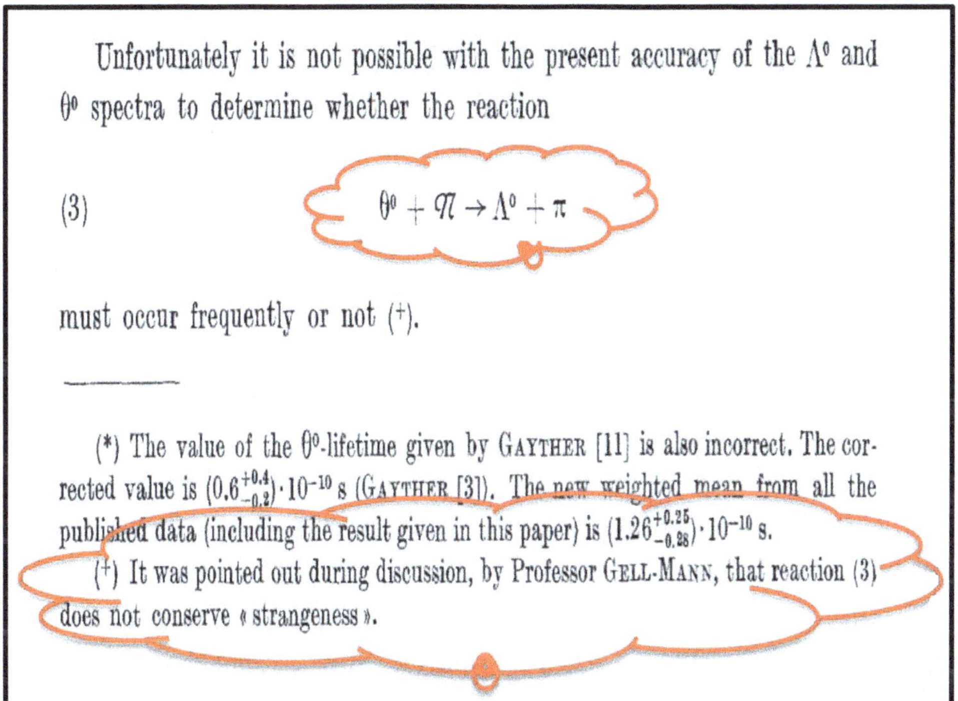

> Unfortunately it is not possible with the present accuracy of the Λ^0 and θ^0 spectra to determine whether the reaction
>
> (3) $\qquad \theta^0 + \mathcal{n} \rightarrow \Lambda^0 + \pi$
>
> must occur frequently or not (†).
>
> ———
>
> (*) The value of the θ^0-lifetime given by GAYTHER [11] is also incorrect. The corrected value is $(0.6^{+0.4}_{-0.2}) \cdot 10^{-10}$ s (GAYTHER [3]). The new weighted mean from all the published data (including the result given in this paper) is $(1.26^{+0.25}_{-0.28}) \cdot 10^{-10}$ s.
> (†) It was pointed out during discussion, by Professor GELL-MANN, that reaction (3) does not conserve « strangeness ».

Figure 1

As reported by G.D. James, it was pointed out by Gell-Mann that reaction (3) did not conserve the strangeness quantum number.

This was a great event for my career. Professor Butler, the co-discoverer of the V–particles, was the Chair of the Session (**Photo 4**).

(Picture courtesy of the Imperial College London)
Back row left:
Prof. P.T. Matthews, Prof. M. Blackman, Prof. J. McGee, Prof. Sir J. Mason
Front row left:
Prof. A. Salam, Prof. C.C. Butler, Lord P.M.S. Blackett, Prof. W.D. Wright, Prof. H. Eliott

Photo 4

G.D. James did not answer Gell-Mann's objection. In my broken English I said that at production (10^{-23} sec) the θ^0 has s = **+1** but, according to the most recent developments of our understanding of the physics of strange particles, the heavy meson θ^0, later becomes a mixture of a meson with s = **+1**, θ^0, and an antimeson, $\bar{\theta}^0$, with s = **−1**: ($\theta^0 \pm \bar{\theta}^0$). Due to my broken English, no-one really understood what I was saying. The chairman realized that I was defending the "discovery" and said: "Could you please come to the blackboard." This gave me the chance to be understood by everybody since I was using equations not words, and the conclusion was me writing the correct reaction

$$\bar{\theta}^0 + n \to \Lambda^0 + \pi,$$

where the "strangeness" quantum number s was indeed conserved, since it was -1 for the $\bar{\theta}^0$ and for Λ^0, and **zero** for the other two particles n and π.

In just a few minutes, from being a totally unknown fellow, I became the most famous physicist of the conference. Few words on the discussion. "The only difficulty" I said "is that there was no experimental evidence for the production of heavy mesons with positive and negative strangeness." This was a top-priority challenge in order to overcome the Enrico Fermi criticism against the "strangeness" quantum number. In fact, the introduction of this new quantum number in the description of the fundamental properties of the nucleon and of the meson (pion) was destroying their spectacular property: the nucleon is a fermion in Lorentz-Space-Time and is also a fermion in the intrinsic-space (isospin) (see "*Fermions and Bosons*", Appendix 4); the meson (pion) is a boson in Lorentz-Space-Time and also a boson in the intrinsic-space (isospin). The isospin-space was the new frontier of physics and its close link with the Lorentz-Space-Time was extremely interesting. This link gave the result that fermions must be isofermions and bosons must be isobosons.

This was the status of particle physics at that time, as recalled by Gell-Mann in his 80th Anniversary Celebration Lecture [6]. The physicists of that time really liked this spectacular property of mesons and nucleons, since it allowed the establishment of a very interesting connection between the intrinsic property (fermionic and bosonic) in Lorentz-Space-Time and in the isospin-space. No one had the slightest idea that many years later the "no-go" Theorem of Coleman and Mandula would have been discovered [7] and that with the discovery of Supersymmetry [8] (see "*Supersymmetry and Superworld*", Appendix 5), fermions and bosons would have been put on an equal basis, thus opening the new horizon for the existence of the Superworld [9]. Let me recall some of the theoretical achievements which originated in the discovery of Supersymmetry, i.e. the discovery of Supergravity [10] and of the M–theory [11]. Unfortunately, the Superworld is still missing direct experimental evidence for its existence [12].

Going back to 1947, in addition to the "destructive" power of the so much wanted connection between the Lorentz-Space-Time and the intrinsic space properties, there was another fundamental objection for the existence of the "strangeness" quantum number to be additively conserved. If a new quantum number, called "strangeness", really exists, it cannot be **+1** for the heavy meson θ^0 and **−1** for the neutral baryon Λ^0. The two opposite values of a quantum number **±1** must belong to the same particle/antiparticle state.

It cannot be (**+1**) for a particle which is a heavy meson, and (**−1**) for another particle which is a baryon. This assignment was based on the experimental discovery of the associated production of two V^0s, identified to be a baryonic event ($\Lambda^0 \to p\pi^-$) and a heavy mesonic event ($\theta^0 \to \pi^+\pi^-$).

What was needed was the experimental proof of pair production of heavy mesons with positive and negative strangeness, as it happens with the pair production of (e^+e^-) discovered by Blackett in 1932. This was my conclusion of the discussion session (**Photo 5**).

Photo 5: A.Z. and Murray Gell-Mann talking about heavy mesons with positive and negative strangeness.

After the 1955 conference, Professor Blackett asked me if I was interested in participating in a competition to join his group. This is how I became the youngest member of his group. My top priority was to study **the photos**, taken by the Blackett group at the Jungfraujoch Lab (3,580 meters a.s.l.).

II-2 – HOW IT COULD HAPPEN THAT – FEW YEARS AFTER CERN WAS ESTABLISHED – THE DISCOVERY NEEDED FOR THE EXISTENCE OF THE "STRANGE" CHARGE WAS OBTAINED

It took me two years (from 1955 to 1957) to find two photos, where the pair production of heavy mesons (whose name from θ became K), with positive and negative strangeness, was experimentally proved to exist [5]. This was the first discovery at CERN, founded in 1954.

The front page of the paper [5] is reproduced in **Figure 2**.

Examples of the Production of $(K^0, \overline{K}^0)$ and $(K^+, \overline{K}^0)$ Pairs of Heavy Mesons.

W. A. COOPER, H. FILTHUTH, J. A. NEWTH, G. PETRUCCI, R. A. SALMERON and A. ZICHICHI

C.E.R.N. - Geneva

(ricevuto il 14 Gennaio 1957)

Summary. — Two simple nuclear interactions that produce pairs of K-mesons are described and discussed. They are interpreted as examples of the processes $n + p \rightarrow K^0 + \overline{K}^0 + n + p$ and $n + p \rightarrow K^+ + \overline{K}^0 + n + n$ where $\overline{K}^0$ is the anti-particle of the K^0-meson.

Conclusion.
We interpret our observations, therefore, as being examples of $(K^0, \overline{K}^0)$ and $(K^+, \overline{K}^0)$ pairs produced in elementary neutron-proton interactions.

Figure 2

We will see in Chapter II-3 that Blackett was the first physicist who started to work with Wilson-chambers electronically triggered.

In the Jungfraujoch photos, there was a lot of physics to be found. In the photos taken at the Cervinia Lab (3,480 meters a.s.l.) it was very rare to find an interaction that produced many particles. How do you explain this?

Professor Blackett was the author of the *Electronic Trigger System with threshold Energy* at the level of at least ten GeV:

$$E_{Rays}^{Cosmic}\ Interaction\ \geq 10\ GeV.$$

The Cervinia Trigger was at an energy level ten times lower $E \gtrsim 1$ GeV.

This is why the photos taken at Cervinia had very little physics in. We will see in Chapter III-3.1 that at CERN also a discovery in the Physics of high precision was achieved thanks to the invention of a new technology able to allow the construction of a series of magnetic fields hundreds times cheaper and hundreds times faster to produce than all know technologies.

When a new laboratory is proposed the target is always along three lines of research: new physics, new high precision measurements of fundamental quantities, and new technological inventions.

This could be achieved in the new institution called CERN, in few years, thanks to Blackett, Bohr and Rabi. They are the authors of the fundamental actions in establishing CERN. They wanted to give this new institution a scientific component, in order to have, from the very beginning, a strong presence of Science in addition to administrative structures to avoid the invasion of bureaucratic powers in a scientific institution. Rabi, whose role was essential since he was the scientific advisor of the USA President, was convinced that this was a fundamental property (**Photo 6**). The official date for the existence of CERN is 1954 but the correct time should be one year before since it is in 1953 that Rabi convinced Dwight Eisenhower, the USA President, that the scientific activity of the new Europe emerging after the 2nd World War had to be encouraged by the USA.

Photo 6: Rabi (Erice 1973) with his former student Norman Ramsey while preparing a lecture on the origin of CERN to celebrate the 20th anniversary.

Blackett offered his cosmic ray group, which was the most powerful in the world. Bohr who was responsible for getting the support of the Scandinavian Countries, considered the scientific activity for CERN a vital condition to start with. The results of the Blackett, Bohr and Rabi "*vital condition*" are in the Figure below (**Figure 3**).

THE RESULTS OF THE BLACKETT, BOHR AND RABI "*VITAL CONDITION*" FOR CERN

1 – **New Physics**: the discovery of pair production of heavy mesons with positive and negative "strangeness" (Chapter II-2).

2 – **High Precision Physics**: the anomalous magnetic moment of the muon (Chapter III-3.1).

3 – **The invention of a new technology** for the construction of polinomial magnetic fields (Chapter III-3.2).

4 – **The 1st proof that CERN could compete** with well established famous Labs and win (Chapter III-3.3).

Figure 3

Before going on to the next chapter a synthesis of the 1947 events, where from all started is given in **Figure 4**.

1947 – The end of my studies at the pre-university level.

1947 Great achievements in Physics

① **Virtual Physics** – Lamb–shift [13]: *unpredicted but needed since it was the simplest example of virtual phenomena discovered by Blackett in 1932 with the "vacuum polarization"* [14, 15, 16]

② **The Nuclear glue** – π–meson [17]: *very much wanted as being the missing nuclear glue*

③ **The V–particles** [18]: *unpredicted with enormous consequences started with the new quantum number called strangeness*

Figure 4

1948 + 5 ≡> **1953** End of my university studies.

1955 ≡> The youngest member of the Blackett group.

1957 ≡> Discovery of pair production of heavy mesons with $s = \pm 1$.

Thanks to Blackett, the same research group was involved successfully in the three great achievements of **1947**: **Virtual Physics, Nuclear Glue and the V–particles**, the most spectacular **being the V–particles**, as we will see in Chapter III-1.

II-3 – AN EXAMPLE OF ETHICS IN OUR SCIENCE

I would like to close this chapter showing the first page, **Figure 5**, of an experiment [14] where the most advanced technology (cloud chamber using electronic triggering) and the most advanced theoretical frontiers, the Dirac equation, with all its consequences [16], were used and correctly quoted.

This paper is an example to those who have forgotten the ethics needed as basic principle in our work: you can never forget to quote those whose original ideas have opened the way to **new horizons**. The example I have been personally involved refers to the search for the third lepton, called $(HL)^{\pm}$ with its neutrino ν_{HL} proposed in all details and experimentally searched for [19–20] at CERN (in $\bar{p}p$ annihilation) and in Frascati with the new (e^+e^-) collider (ADONE), by the CERN-Bologna group [19] but now called $\tau^{\pm}$ and ν_τ, despite its original names $HL^{\pm}$ and ν_{HL}.

> 699
>
> *Some Photographs of the Tracks of Penetrating Radiation.*
>
> By P. M. S. BLACKETT and G. P. S. OCCHIALINI, The Cavendish Laboratory, Cambridge University.
>
> (Communicated by Lord Rutherford, O.M., F.R.S.—Received February 7, 1933.)
>
> [PLATES 21-24.]
>
> 1. *The Experimental Method.*
>
> We have recently developed a method by which the high speed particles associated with penetrating radiation can be made to take their own cloud photographs.* By this means it is possible to obtain these photographs very much more speedily than by the usual method of making expansions at random. For when this latter method is used it is only on a small fraction of the photographs that a track will be found. The average number of photographs required to obtain one track will depend on the size and orientation of the chamber and on the effective time of expansion. The latter is not likely to be more than 1/20 second. From measurements with counters it is known that about 1·5 fast particles fall, from all directions, on 1 sq. cm. per second. Roughly consistent with these figures are the results found with cloud chambers. Skobelzyn† has obtained as many as one track every ten expansions, but in the work of Anderson‡ the number of tracks which were long enough to be suitable for energy measurements was only about 1 in 50 photographs. By our method, tracks are found on 80 per cent. of the photographs. We intend to give a full account of the technique of this method of photography in a separate paper, confining ourselves here to a rough outline only.§
>
> A cloud chamber of diameter 13 cm. and depth 3 cm. is arranged with its plane vertical and two Geiger-Müller counters, each 10 cm. by 2 cm. are placed one above and one below the chamber so that any ray which passes straight through both counters will also pass through the illuminated part of the chamber, fig 2. An alternative arrangement is described on p. 719. The
>
> ---
>
> * 'Nature,' vol. 130, p. 363 (1932).
> † 'C. R. Acad. Sci. Paris,' vol. 195, p. 315 (1932).
> ‡ 'Phys. Rev.,' vol. 41, p. 405 (1932).
> § Mott-Smith and Locher 'Phys. Rev.,' vol. 38, p. 1399 (1931); vol. 39, p. 883 (1932) have found a correlation between the occurrence of these tracks and the discharge of a counter, and Johnson, Fleischer and Street, 'Phys. Rev.,' vol. 40, p. 1048 (1932) have used coincidences to operate the illuminating flash of a continuously working closed chamber.

Figure 5

Here the **new horizon** was the third Family of fundamental particles. A volume whose title is *"The Origin of the Third Family"* [20] has been published in order to point out that we cannot forget the ethical principles of our activity, as given in the Blackett and Occhialini paper [14].

The testimony [20] by C.S. Wu: "*A. Zichichi started his scientific activity with P.M.S. Blackett who wanted this young fellow in his cosmic ray group; I.I. Rabi has been a strong supporter of Nino's projects and activities. The third family has its roots at CERN, where the insight of these two great leaders has been best implemented by A. Zichichi through his more than ten years of dedicated work to the problem of searching for a heavy lepton carrying its own leptonic number and being coupled to its own neutrino*".

The reaction against the violation of scientific ethics in our work is at the origin of the special volume [20] whose front page is reproduced below.

And here comes the testimony of Isidor Isaac Rabi: *"The new-comer can never take away from you the privilege of having been the first to open a new field with your intelligence, imagination and hard work. Do not be afraid to encourage others to pursue your dream. If it becomes real, the community will never forget that you have been the first to open the field."*

This statement (**Photo 7**) is in the Dorothy C. Hodgkin courtyard of the I.I. Rabi Institute at the EMFCSC in Erice.

Photo 7

Photo 8: From left A.Z. and Rabi at CERN (Geneva 1962).

III – BLACKETT AND THE DISCOVERY OF THE SUBNUCLEAR UNIVERSE

III-1 – THE V–PARTICLES: FROM STRANGENESS TO QCD AND ITS "HIDDEN" SIDE PLUS THE "END OF A MYTH"

The developments of the branches of physics generated by the **V–particles** are illustrated in **Figure 6**, where the (θ–ι) puzzle on one side and the proliferation on the other gave enormous consequences. The (θ–τ) puzzle gave the violation (≠) of Parity (P) and of charge conjugation (C) plus flavour mixing and PC breaking (≠).

The proliferation of an enormous number of particles had two important components, static and dynamic.

The static one produced first the SU(3)–flavour $(SU(3)_f)$ (with u d s) and later SU(3)–colour $(SU(3)_c)$ i.e. QCD (Quantum ChromoDynamics).

The great goal of the dynamic component was the discovery of the Effective Energy which gave the Universality Features in all reactions, no matter the nature of the interacting particles.

The statics and the dynamics gave QCD with asymptotic freedom and confinement (see "*Asymptotic Freedom and Confinement*", Appendix 6).

In **Figure 6** there are two important details, the $(\theta-\tau)$ puzzle and the LOY (Lee, Oehme, Yang) contribution to disentangling the flavour mixing from the problem of the CP breaking.

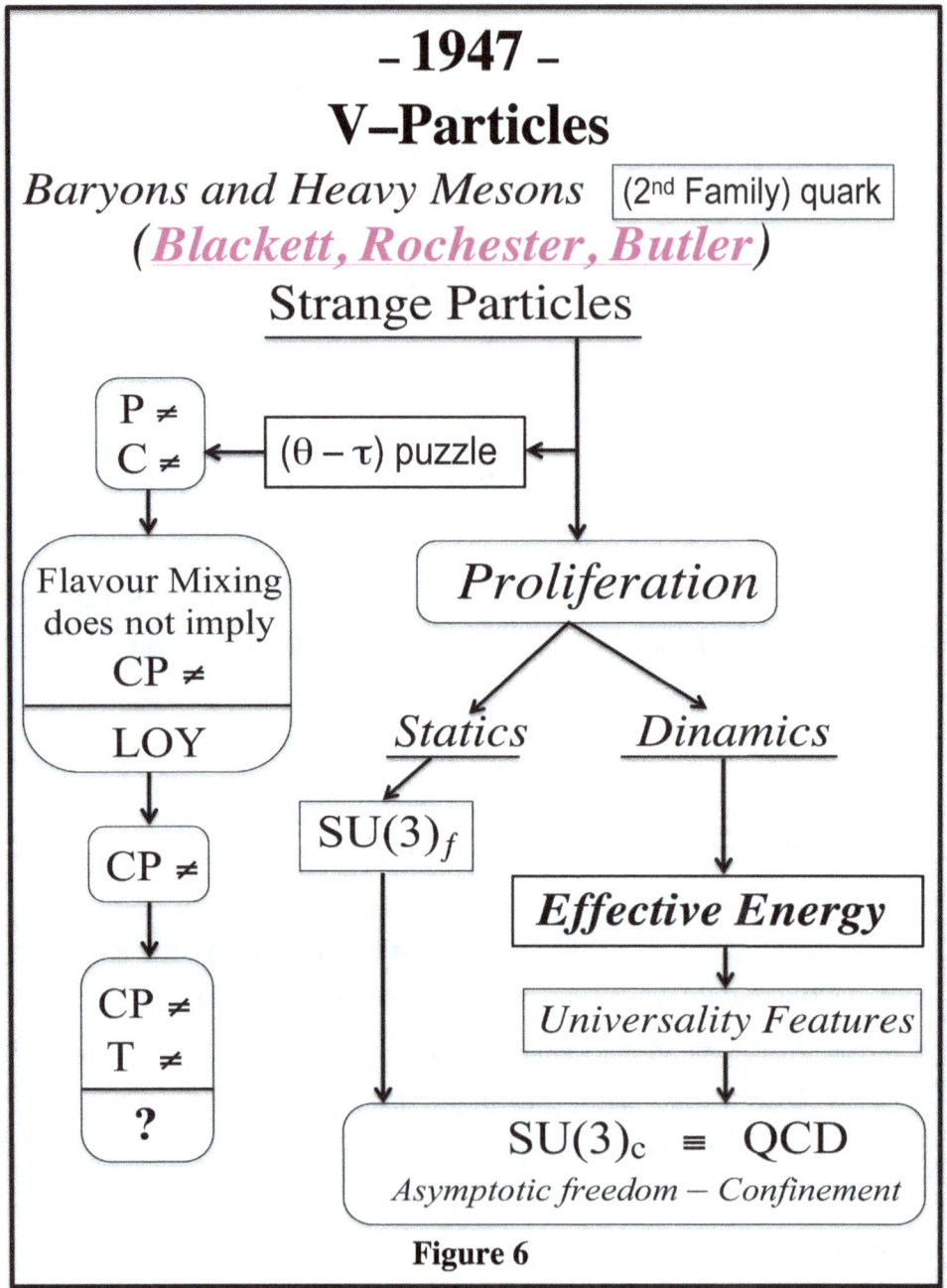

Figure 6

Let us start with the (θ–τ) puzzle [21], which culminated in the discovery of the breaking (≠) of the symmetry operators C and P. The discovery of the non-invariance of these symmetry operators was suggested (1956) in a detailed analysis of all weak processes by T.D. Lee and C.N. Yang [22]; the first experimental evidence was provided by C.S. Wu and collaborators one year later [23]. Nevertheless, there was a lot of confusion in those years. For example the "strangeness mixing", proposed by Gell-Mann and Pais [24], to describe the ($\theta^0\,\bar\theta^0$) pair, brought them to predict the existence of θ_1^0 and θ_2^0, on the basis of the validity of C invariance in weak interactions (see "*Weak Interactions*", Appendix 7). The discovery by Lederman of $\theta_2^0 \to 3\,\pi$ [25] was interpreted as a proof that C invariance holds in weak interactions. With the discovery of C and P breaking, the (θ–τ) **mesons became a unique particle**, the **K–meson**, which split into two components, K_1^0 and K_2^0, each one thought to be an eigenstate of the symmetry operator CP proposed by Landau [26] to replace the two broken P and C invariances. Probably few people knew that, **in 1956, Lee, Ochme and Yang**, (**LOY**) before parity violation was experimentally proved by C.S. Wu, pointed out that the existence of K_2^0 could not be taken as a proof of C invariance, nor as a proof of CP invariance [27]. Lee, Oehme and Yang (LOY) showed that "strangeness mixing" does not imply C invariance as claimed by Gell-Mann and Pais. In fact, even if CP is not valid, K_2^0 would still be there and, in order to prove that "strangeness mixing" is or is not CP invariant, other experiments had to be done in K decay physics. In 1964, it was discovered that CP and T invariances are broken [28], as suggested by LOY [27].

There is an amusing detail of this great discovery [28]. The experiment was not planned to search for the 2 π decay mode of the K_2^0 meson. The aim of the experiment was to check the anomalous regeneration in hydrogen, previously reported by Robert Adair *et al.* [29] and later found [28] to be more than an order of magnitude lower.

The search for the 2 π decay mode of the long-lived K_2^0 was proposed by the CERN-Bologna group at CERN, but rejected because the special experimental hall of the Proton Synchrotron (PS) where the neutral beam was operative, had already been allocated to another group's programme. On the other hand, the CERN-Bologna group was already engaged with the PAPLEP (Proton AntiProton Annihilation into LEpton Pairs) experiment to search for the production of the third lepton through the (eμ) final state produced in ($\bar{p}p$) annihilation [19–20], using the CERN-PS beam which was next to the neutral beam we wanted for the $K_2^0 \rightarrow 2\pi$ search.

I was told by the CERN Research Director of the time "give other people the chance", when trying to convince him that the existence of the long lived K_2^0 was not proof of CP invariance as shown by LOY in 1957 [27], therefore the search for the $K_2^0 \rightarrow 2\pi$ decay mode violating CP invariance was not in contradiction with the existence of the long lived K_2^0 meson. It would have been too much to give two PS beams to the same group, he told me later. Moreover we were not proposing to check the anomalous regeneration in hydrogen (a proposal considered very interesting). Our aim was to follow the LOY theoretical deep remark and check if CP was really valid.

Professor Blackett was convinced that large institutions like CERN had to solve the problem of decision-making in frontier physics where high scientific responsibility could be given to the wrong fellows.

The flavour mixing problem and its CP invariance or non-invariance is extremely topical even today with many experiments being planned in order to understand the basic distinction between "flavour mixing" and CP invariance, for all flavours. Very few people could imagine the relevance of this problem in those days. The fact that the authors of the basic distinction between "flavour mixing" and CP invariance are LOY has been completely forgotten.

How and why the quark flavours (up, charm, top, or u, c, t) and (down, strange, bottom, or d, s, b) mix, and why this mixing is linked with the breaking of CP has no theoretical understanding, so far. This is the meaning of the question mark in **Figure 6**. All we can do is to measure the various flavour mixings and CP breakings.

Flavour mixing appears to be active also in the lepton sector. Sooner or later, these problems need to be understood.

The other chain of consequences originated by the existence of the V–particles was, as illustrated in **Figure 6**, the proliferation of mesons and baryons with two branches: "statics" and "dynamics". Let us add few words to what has already been said.

The "static" proliferation gave rise, first to an order of magnitude reduction of the mesonic and baryonic states via the eight-fold way of Gell-Mann and Ne'eman [30], and then to the "flavour" ("f") global symmetry $SU(3)_f$ based on the existence of three quark flavours: u, d, s [32].

We know that the number of quark flavours is six: u d c s t b, thus the "static" reduction of proliferation via $SU(3)_{f\,\equiv\,uds}$ was an illusion.

The unexpected consequence of $SU(3)_f$ was its contribution to open the way towards the existence of the "colour" (c) $SU(3)_c$. It is in fact the notion that two baryons

$$\left(N^*\right)^{++}_{3/2,\,3/2} \quad \text{and} \quad \Omega^-,$$

had to be fermions, but appeared to be perfectly symmetric in their quark composition [33], that prompted the idea for the existence of a new intrinsic quantum number [34–35].

This V–chain of consequences (with the other chain in the "virtual physics" discussed in Chapter III-3), led to the discovery [36] of $SU(3)_c$, whose final goal was Quantum ChromoDynamics (QCD) [37], the last point in **Figure 6** $SU(3)_c \equiv QCD$. This last point has attracted the attention of the physics community during many decades. In fact **QCD** had also a "hidden" side which needed the discovery of the "Effective Energy" in order to be solved. Let us try to illustrate in few lines these many decades of activities.

We start with **the "hidden" side of QCD**. The "hidden" side of QCD [38], is due to the proliferation in the "dynamics": how it can be that the unique fundamental force acting among its two very simple basic components, quarks and gluons, produces such a variety of final states as those observed in **strong**, **electromagnetic** (EM) and **weak** interactions.

It is the **Effective Energy** [39] which makes it possible to overcome this "hidden" trouble of QCD, as we shall now see. First the definition of "hidden" with the famous Vladimir N. Gribov's remark: *"In the physics community there was a sort of gentlemen's agreement: please do not speak about results in contrast with the so much searched for gauge interaction to describe hadronic phenomena. These "hidden" results were the hadronic systems produced in the interactions between pairs of hadrons; they were all different. Each pair of interacting particles, when producing systems consisting of many hadronic particles, had its own final state. No-one knew how to settle this flagrant contradiction. I wish I had the idea of the "effective energy."*

This remark and a few other contributions are in a volume on *"The Creation of Quantum ChromoDynamics and the Effective Energy"* edited by Lev N. Lipatov [38]. Here is a synthesis on the status of strong interactions by Victor Weisskopf: *"The physics of strong interactions whose "elementary particles" were of baryonic and mesonic nature was characterised by two classes of phenomena, both showing a very large number of different varieties. One was of static nature: i.e. the enormous number of mesons and baryons. The* *other was of dynamic nature: the enormous number of different final states produced by different pairs of interacting particles, such as (πp), (pp), ($\bar{p}p$), (Kp) in addition to (e^+e^-), (νp), (μp), (ep) etc.. These multihadronic final states were all measured to be different. Why do the multihadronic final states depend on the nature of the colliding pair?"*

Lev N. Lipatov: *"In order to be able to "predict" Universality Features, QCD needs to be formulated in such a way that both perturbative and non-perturbative effects are included simultaneously."*

Gerardus 't Hooft: *"Theoreticians were unable to prescribe what experimentalists had to look for to establish the universal nature of these final interactions. The experimental results were discouraging; scattering experiments yielded different final states for each pair of interacting particles. So it happened that these aspects of QCD had to wait until experimentalists themselves came with the right idea [38]. The showers come with what is now called an "effective energy", and, in terms of this quantity, universality could be established [39]."*

Few more words on the **Effective Energy** (see *"The Effective Energy"*, Appendix 8).

The proliferation in the "dynamic" sector was the multitude of final states produced by pairs of interacting particles, in Strong, Electromagnetic (EM) and Weak processes:

Strong	EM	Weak
π p	γ p	ν p
K p	e p	
p p	μ p	
p n	e^+e^-	
$\bar{p}$ p		

It is the introduction of the Effective Energy which allows us to put all the different final states on the same basis.

This basis is the quantities measured in the multihadronic final states:
- (i) the average charged multiplicity: $<n_{ch}>$;
- (ii) the fractional energy distribution: $d\sigma/dx_i$;
- (iii) the transverse momentum distribution: $d\sigma/dp_{T_i}$; etc. ..... .

If the Effective Energy does not change, these quantities do not change, no matter what the pair of interacting particles in the initial state is. The result is the universality features measured in all multihadronic final states.

The universality features are a QCD non-perturbative effect.

The first and basic step in this long "non-perturbative" QCD trip is the introduction of the Effective Energy.

This **new quantity** came about by studying (pp) interactions at the CERN-ISR, the most powerful (pp) collider in the world, constructed under the CERN-Directorship of Victor Weisskopf. Here are his words: *"One of my contributions to CERN was the decision to construct the first proton-proton collider of the world, the Intersecting Storage Rings (ISR). The ISR allowed to reach the highest collision energy in the interaction between protons. I am therefore very happy that it is thanks to a series of experiments performed by A. Zichichi and collaborators with the ISR that the multitude of final states produced when particles interact strongly, electromagnetically and weakly can be put on the same basis"* [38].

It was proved that the set of final states produced at the ISR nominal energy of

$$62 \text{ GeV}$$

consisted of a series of final states, each one having a different Effective Energy going from the lowest (just a few GeV) up to the highest ($\simeq 30$ GeV) values allowed by our experimental set-up. The "nominal" ISR energy (62 GeV) had to be corrected event-by-event.

The final states had properties like those produced in (e^+e^-) annihilation at the same "effective" ISR energies.

A quotation from the first paper where these results are reported is shown in **Figure 7**.

Introduction of the "Effective Energy"

EVIDENCE OF THE SAME MULTIPARTICLE PRODUCTION MECHANISM IN p-p COLLISIONS AS IN e^+e^- ANNIHILATION

M. Basile, G. Cara Romeo, L. Cifarelli, A. Contin, G. D'Alì, P. Di Cesare, B. Esposito, P. Giusti, T. Massam, F. Palmonari, G. Sartorelli, G. Valenti and A. Zichichi

Physics Letters 92B, 367 (1980).

"The agreement between the momentum distributions obtained in e^+e^- annihilation and in pp collisions suggests that the mechanism for transforming energy into particles in these two processes, so far considered very different, must be the same".

Figure 7: The first paper where the Effective Energy was introduced in the study of high energy (pp) interactions at the nominal ISR energy (62 GeV).

Here is Gribov again: *"When I read the paper "Evidence of the same multiparticle production mechanism in pp collisions as in e^+e^- annihilation" by A. Zichichi and collaborators, working with the pp ISR collider at CERN, I* *realized that something very interesting had been found. In fact the introduction of the 'effective energy' in the analysis of pp collisions at CERN's ISR gave a totally unexpected result"* [38].

And now Gabriele Veneziano: *"This work was very timely, as several difficulties were piling up when confronting QCD with experimental data (e.g. the η–η' problem). These difficulties contrasted with its initial successes, when a new intrinsic degree of freedom (colour) was introduced to explain how totally symmetric SU(3)-flavour states, the well known baryons*

$$(N^*)^{++}_{3/2,\ 3/2} \quad \text{and} \quad \Omega^-,$$

could be antisymmetric in the constituent fermions. The elementary constituents of all mesons and baryons thus needed to be interacting quarks and gluons" [38].

In order to check the validity of the Effective Energy down to the lowest value, the ISR collider was used at its lowest nominal energy:

$$(\sqrt{s})_{pp} = 30 \text{ GeV}.$$

This allowed a set of very low Effective Energies to be obtained using purely hadronic interactions. It was shown (**Figure 8**) that the fractional energies of the secondary particles produced in (pp) collisions had the same distribution as those produced in (e^+e^-) annihilation, at the same Effective Energy.

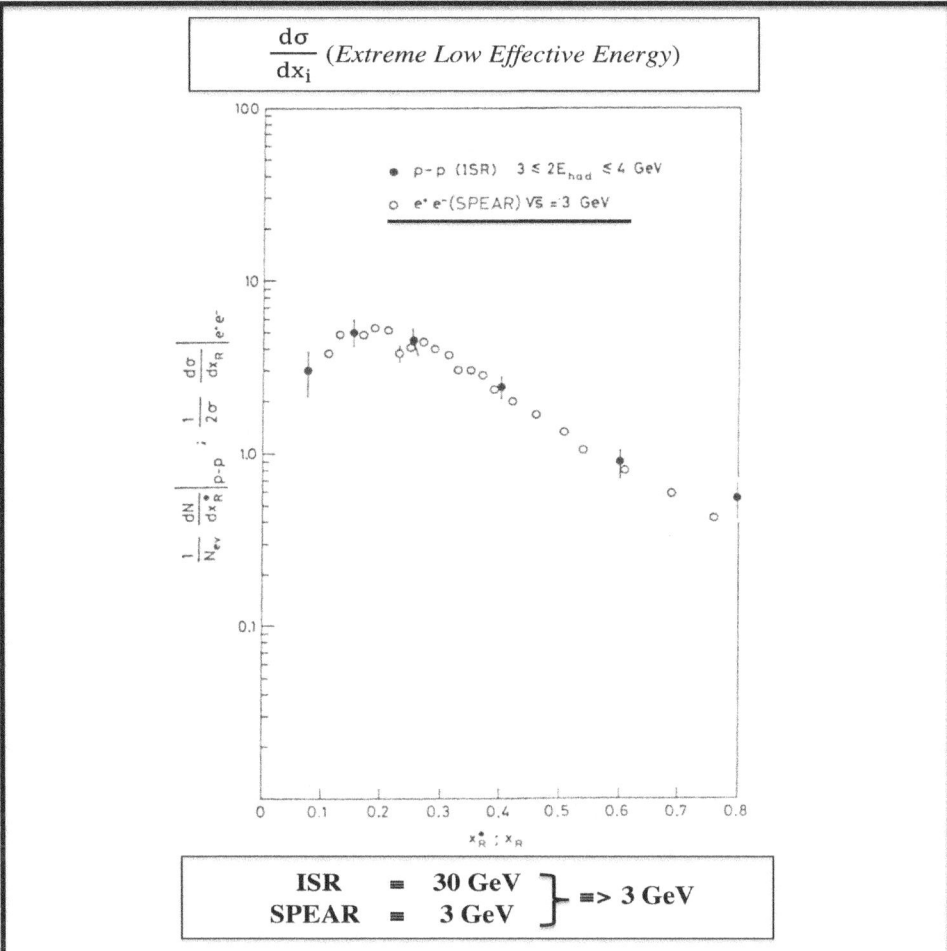

Figure 8: The inclusive single-particle fractional momentum distributions $(1/N_{ev}) \times (dN/dx_R)$ in the interval $3 \text{ GeV} \leq 2E^{had} \leq 4 \text{ GeV}$ obtained from data at the ISR-nominal energy, $\sqrt{s} = 30$ GeV. Also shown are data from MARK I at SPEAR.

During many decades the following processes were considered basically different:

1. Low transverse momentum, P_T, strong interactions.
2. High transverse momentum, P_T, strong interactions.
3. Deep Inelastic Scattering (DIS) in weak (νp) and electromagnetic (ep) plus (μp) interactions.
4. (e^+e^-) annihilation.
5. ($\bar{p}p$) annihilation.

An interesting detail: the low P_T interactions represent, in the field of strong forces, the overwhelming majority of events. There was a sort of myth: only high P_T hadronic processes had to be compared with DIS, as shown in **Figure 9**.

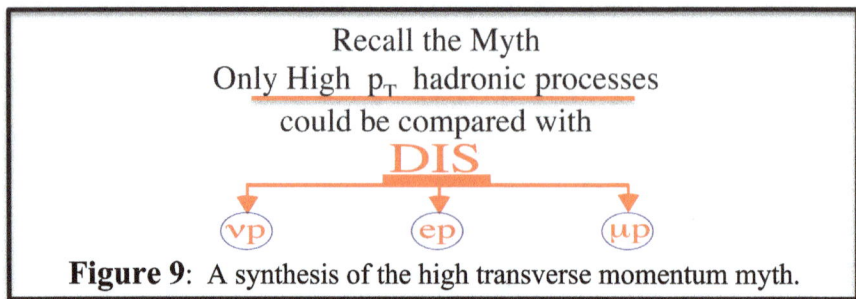

Figure 9: A synthesis of the high transverse momentum myth.

The introduction of the Effective Energy gave the result that multihadronic final states produced in very many different processes are the same, provided the Effective Energy is the same. Details can be found in the volume whose cover page is in **Figure 10** [38].

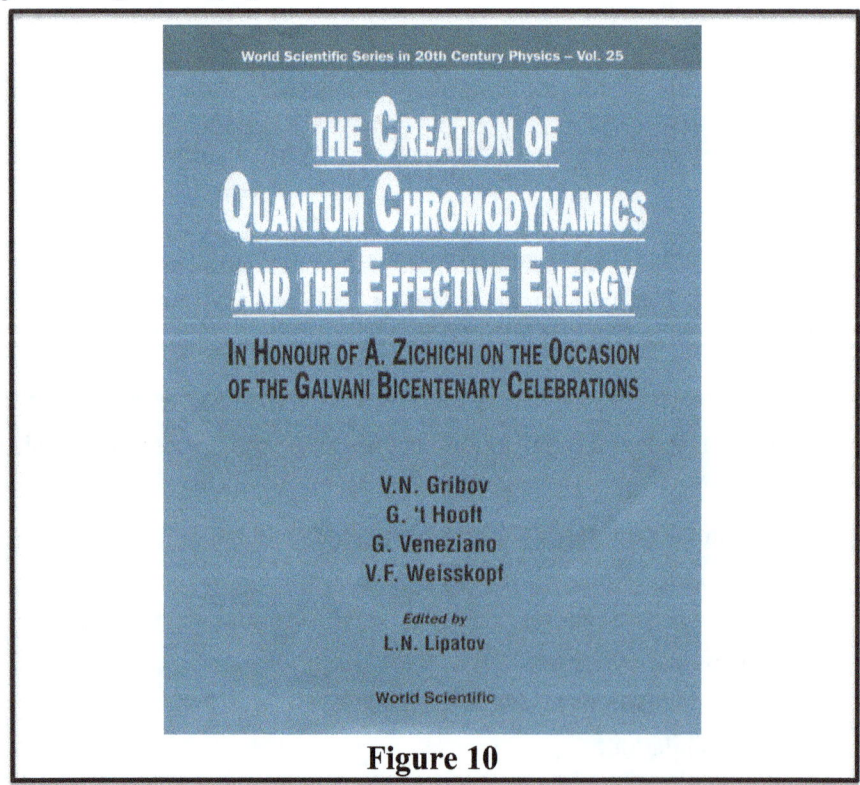

Gabriele Veneziano: *"Paper [39] (**Figure 7**) was followed by a series of other results, all confirming the validity of this discovery. It was indeed a great surprise to see that introducing the effective energy, i.e. the energy effectively available in each reaction for the production of secondary particles, allowed to expose universal features even when the incoming particles were not hadronic"*, [38], as reported in **Figures 9** and **10**.

And now the final theoretical trouble affecting QCD: confinement.

From QCD, it is not possible to predict confinement (see *"Asymptotic Freedom and Confinement"*, Appendix 6). But our ISR results [40] prove that, at high energy, confinement holds.

So, QCD must produce:

    (i) Asymptotic freedom, in order to explain SLAC scaling, and

    (ii) Confinement, in order to explain the ISR results of no quarks at high energy.

Here is the way Gerardus 't Hooft summarized the status of scaling and confinement. G. 't Hooft: *"In spite of their beauty, these ideas were taken with skepticism. The explanation of quark confinement was not considered to be adequate. This skepticism encouraged a new series of experiments at CERN, where the highest energy pp collider, ISR, allowed to check if quarks remained confined at these energies [40]. It seemed to be a miracle that quarks can move freely at high energies and yet they stick together inseparably inside the hadrons"* [38].

It was at the EPS Palermo conference, that G. 't Hooft presented his way to explain confinement [41]. Assume that, in QCD, in addition to the quarks, there are scalar particles with imaginary masses and "colour-magnetic" charges.

If this happens, the colour-magnetic-QCD charges, monopoles, condense, thus providing permanent confinement for the QCD-colour-electric charges, i.e. the quarks. The 't Hooft model is the only theoretical model on the "hidden side of QCD" allowing the existence of confinement.

I would like to close this series of memories about my activity by remembering the teaching of my great master P.M.S. Blackett: "*Never abandon a problem which has attracted your interest. The originality of an idea must be translated into the implementation of an original experiment*". This is why in the volume of **Figure 10** on the **End of a Myth** (see "*The End of a Myth*", Appendix 9) it is written: "*The root of this new approach to the study of hadronic interactions goes back a long time to a proposal by the CERN-Bologna group*" started before 1969 and never abandoned.

III-2 – THE NUCLEAR GLUE: FROM THE π–MESON TO THE THIRD FAMILY PLUS THE GLUON JETS AND THE INSTANTONS

The π discovery provided the "glue" for the nuclear forces and this was great. Nevertheless the formidable role of the π–meson turned out to be in the field of lepton physics.

As illustrated in **Figure 11**, this discovery was based on the observation of the complete decay-chain-reaction

(1) $$\pi \rightarrow \mu \rightarrow e \ ,$$

which allowed one to understand the real nature of the cosmic ray "meson" first observed in 1936 by Anderson and Neddermeyer [42]: this "meson" was not the π but the µ.

Despite the war, three Italian physicists, Conversi, Pancini and Piccioni, were able to perform an experiment in Italy where the negative component of the cosmic ray "meson", thought to be the π, was stopped and observed to decay as if it had no nuclear interaction with matter [43]: in fact the "meson" stopped and observed to decay was not the π but the µ.

The cosmic ray "meson" studied by Conversi, Pancini and Piccioni was not Yukawa's meson, π, but its decay product, the μ, as finally and very clearly proved by the above mentioned complete decay-chain-reaction (1) (see also *"The Yukawa's meson: the first example of the nuclear glue"*, Appendix 10).

It was G. Puppi who first pointed out that the Fermi couplings of different weak processes, including the muon, were the same [44], within an order of magnitude. The consequences of the Puppi-"triangle" and other details are in References [45–48].

The identification of the muon as a particle deprived of the strong force opened the problem of the second lepton "μ".

If it were not for the π–meson, no-one would have known the existence of this second lepton.

As we shall see later, the accurate measurements of its properties, electromagnetic $(g-2)_\mu$ [49–53] and weak (τ_μ) [54] plus polarization [55], are at the origin of the idea that an experimental search for a heavier lepton [19–20], not having the privilege of being produced by another "ad-hoc" meson, had to be implemented.

The existence of the Vs discovered by the Blackett group implied the existence of the nuclear glue (π–meson), with all the consequences illustrated in **Figure 11**, where the π → μ → e chain opened the problem for the existence of the third lepton (HL) and the existence of a third Family.

In fact, the reason why there were so many μ, also called "heavy electron", was – as mentioned earlier – because a very light meson, the π, exists; this very light meson is produced in strong interactions.

If a heavier meson (in the GeV mass range) existed it would be produced in strong interactions and would decay into a heavy lepton (HL) which would have escaped all experimental observations.

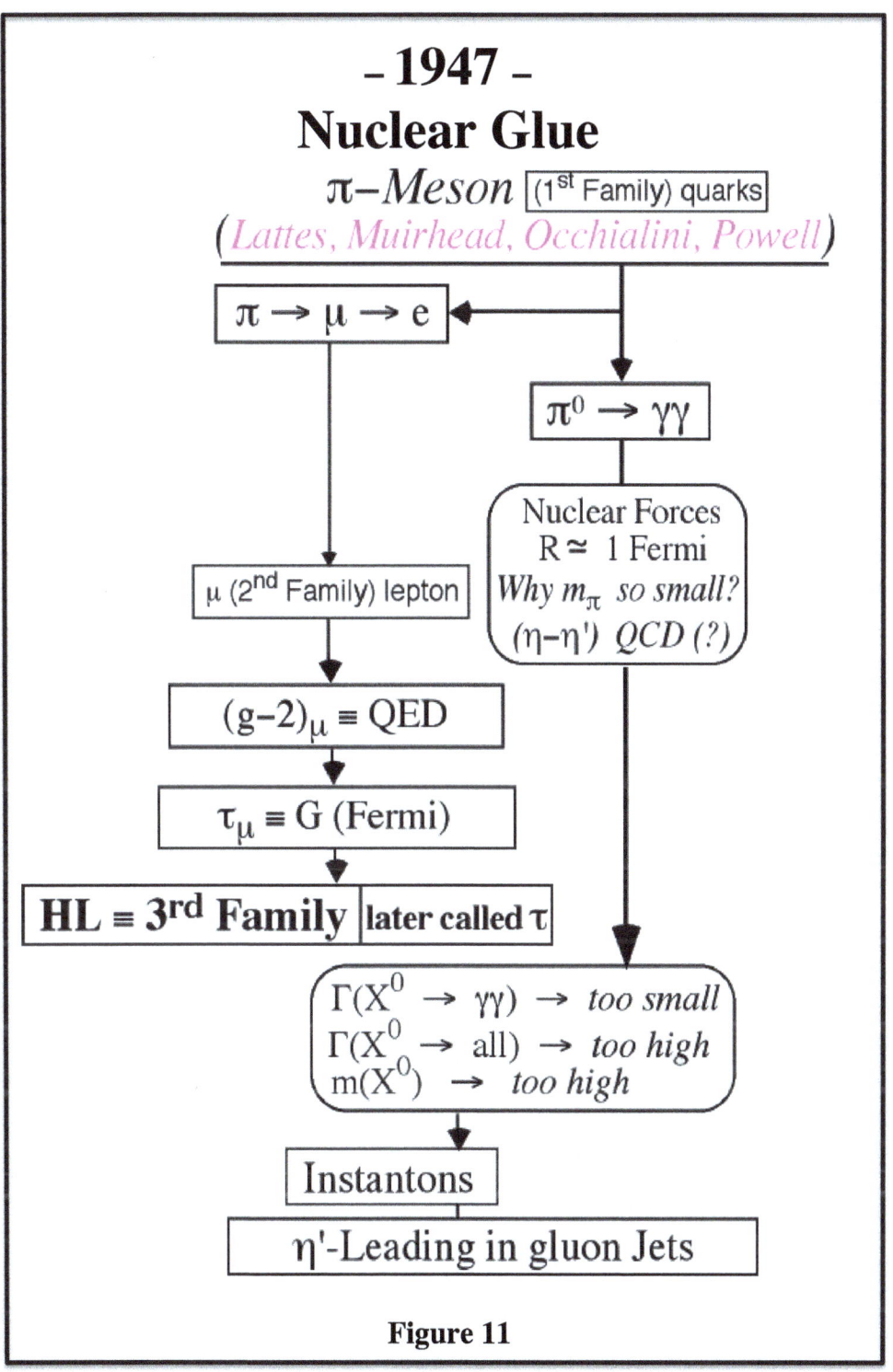

Figure 11

When the search for the third lepton was carried out at CERN and at Frascati [56] its name was HL. Now, contrary to the ethical principle of our activity taught by Blackett and Occhialini in their paper [14] reported in Chapter II, it is called τ.

And that is not all.

The problem of understanding the "nuclear glue", which is the quantum state of the nuclear forces, with the very light π–meson, was followed by the discovery of other mesons, called η and X^0. The η–meson was nearly four times heavier than the π–meson, despite being pseudoscalar like the π. Later another meson called X^0 was discovered with a mass heavier than the nucleon. Its spin-parity was thought to be 2^-. If this was correct no problem: it was not another pseudoscalar meson.

The discovery of the 2γ decay mode of the X^0-meson [57–58] gave a strong support to its pseudoscalar nature. However, its composition in terms of a quark-antiquark pair remained unclear. In fact, if a meson is made of a $(q\bar{q})$ pair, since quarks carry electric charges, the 2γ decay must be easily allowed. The branching ratios of the 2γ decay mode of the two heavy pseudoscalar mesons, η and X^0, were quite different and the absolute widths of the three pseudoscalar mesons, $\Gamma(\pi^0 \to \gamma\gamma)$, $\Gamma(\eta^0 \to \gamma\gamma)$ and $\Gamma(X^0 \to \gamma\gamma)$ did not follow the theoretical expectations.

Another difficulty was the X^0-mass. If the X^0-meson had to follow the Gell-Mann-Okubo (quadratic) mass formula, the mixing angle needed for these two pseudoscalar mesons had to be very small because the X^0-mass is nearly one GeV, compared with the $\simeq 0.5$ GeV η^0-mass. This mixing, when compared with the $(\omega$-$\phi)$ mixing, also measured [59] to be large (as expected), was the smallest known in all meson physics [60].

Now the pseudoscalar nature of the X^0-meson is established and this meson is designated with the symbol η'. The notation now used is:

i) η^8, to indicate the 8th component of the $(q\bar{q})$ content of the pseudoscalar meson $SU(3)_f$ multiplet.

ii) η^0, to indicate the $SU(3)_f$ singlet component of the pseudoscalar $(q\bar{q})$ system.

These two components, η^8 and η^0, are not enough to describe the η' composition. In fact, we think we know the reason why the (η-η') mixing angle is so anomalously small, namely the large gluonic content of the η'.

In QCD, the η and η' have played a decisive role. In the early days there was the so-called η-problem. The theory appeared to demand a pseudoscalar η as an isosinglet made of non-strange quarks, and an η' as an $(\bar{s}s)$ state. Consequently the η-meson had to be close to the pion mass and the η' mass had to be near the K mass. The fact that experiments gave a quite different picture was attributed to the ABJ anomaly [61–62] by Gell-Mann, Fritzsch and Leutwyler [36] and finally explained as an **instanton** effect by G. 't Hooft [63]. Instantons induce a strong coupling between the η' and the two gluons state, and give this state a high mass, both of which may explain why the total width of the η' is so much bigger than that of the η. And consequently why the $\gamma\gamma$ branching ratio of the η' is so small [58].

Concerning experiments, for a number of years many attempts were made to find out the gluonic content of the η', for example via a comparative study of the radiative decays of the (J/ψ) into η and η'. However, all the methods adopted were for a long time based on indirect evidence. Only many years later the first direct evidence for a strong gluonic composition of the η'-meson was discovered [64]. Here are the steps which produced the goal for the strong gluonic composition of the η'-meson. If the η' has a strong gluon pair component, we should expect to see a typical QCD non-perturbative effect: the leading production in gluon-induced jets. In fact the leading effect had been observed in all hadronic processes where some conserved quantum numbers flow from the initial to the final state. If the gluon quantum numbers flow from an initial state made of two gluons into a final state made of η', this meson should be produced in a leading mode when the initial state is made of gluons. This is exactly the effect which was reported [64] for the production of the η'-mesons in gluon-induced jets. The leading effect in η'–production in gluon jets is shown as the last goal of the nuclear glue in **Figure 11**.

If Professor Blackett was with us he would see that the synthesis in **Figure 11** is a proof of his way of thinking: when an unexpected discovery comes in, no one is able to imagine its consequences.

III-3 – VIRTUAL PHYSICS: FROM THE "VACUUM POLARIZATION" TO THE GRAND UNIFICATION OF ALL FORCES AND THE GAP

The discovery of the Lamb-shift was the simplest example of Virtual Physics. Thanks to "virtual physics" we can now study the Unification of the Fundamental Forces of Nature, the famous Grand Unified Theory, GUT. The "virtual physics" is now the frontier of High Energy Physics. All this started with the vacuum polarization effect, based on the experimental discovery of Blackett and Occhialini in 1932 (see Chapter II and **Figure 4**).

In 1932 Blackett with his young fellow Occhialini, discovered the associated production of the (e^+e^-) pair, thus giving experimental support to the existence of the so called "vacuum polarization" processes theoretically envisaged by Dirac thanks to the discovery of his equation (1929). The existence of "virtual physics" started with this discovery (see "*Virtual Physics and the Annihilation Processes*", Appendix 11). We will see in this section how the pair production (e^+e^-) gave rise to the effect called "vacuum polarization".

There is a very amusing detail. Before 1947, no one was able to imagine the Lamb-shift, a "virtual physics" effect much more simple than the "vacuum polarization". We all work thanks to the existence of the enormous number of "virtual physics" phenomena. I will try to explain what it is.

Imagine the most powerful technology. No matter how powerful and precise it is, this technology will be unable to "directly" observe "virtual phenomena". You could say: these phenomena do not exist.

Here was the great novelty during those times. A virtual phenomenon produces rigorously theoretically computable and experimentally observable "effects". Example: the so-called "anomalous" magnetic moment of the "heavy lepton" called "muon", its symbol being $(g-2)_\mu$ (see Chapter III-3.1).

Weisskopf in 1934 was able to calculate that the "vacuum polarization" effect in hydrogen (**Graphic 1**) is negative and very small: $\simeq$ **− 27 Mc/sec**.

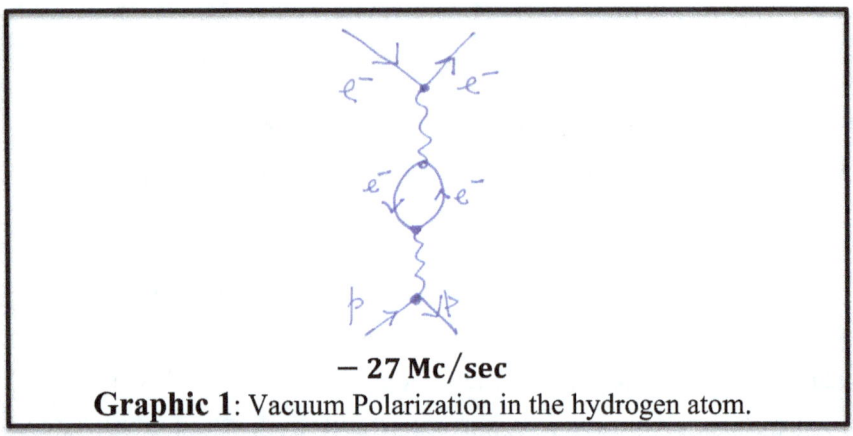

− 27 Mc/sec
Graphic 1: Vacuum Polarization in the hydrogen atom.

The hydrogen atom is the simplest atom of the Mendeleev Table. Its nucleus is made of only one particle: the proton (p). And only one electron is in the atomic cloud around the nucleus. I will now try to illustrate how the pair production of an electron (e⁻) plus its antiparticle (e⁺) predicted by Dirac (**Graphic 2**) and discovered by Blackett and Occhialini [14] gave rise to the virtual effect called "vacuum polarization".

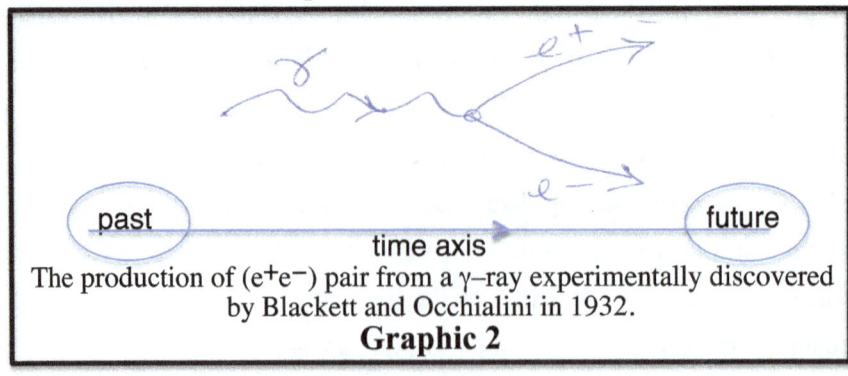

The production of (e⁺e⁻) pair from a γ–ray experimentally discovered by Blackett and Occhialini in 1932.
Graphic 2

The Dirac mathematical formalism says that an electron (e⁻) going from past to future is equivalent to its antiparticle (e⁺) "virtually" going in the opposite time direction, from future to past. From now on particles or antiparticles going from future to past are drawn by dotted lines (– – –).

A photon going from future to past is an antiphoton. Photon and antiphoton are both eigenstates of the C-operator discovered by Herman Weyl and there is no problem: photons and antiphotons are equivalent. The production process of **Graphic 2** is equivalent to the following sequence of events illustrated in **Graphic 3**. A "virtual" electron goes from future to past and is represented by a broken line (– – –). This "virtual" electron going from future to past emits a photon. From this moment the electron is going from past to future and is represented by a full line (———). Thus the diagram in **Graphic 2** is equivalent, according to the mathematical formalism of Dirac, to the diagram in **Graphic 3**. Notice that photon and antiphotons are equivalent and the arrow is therefore omitted.

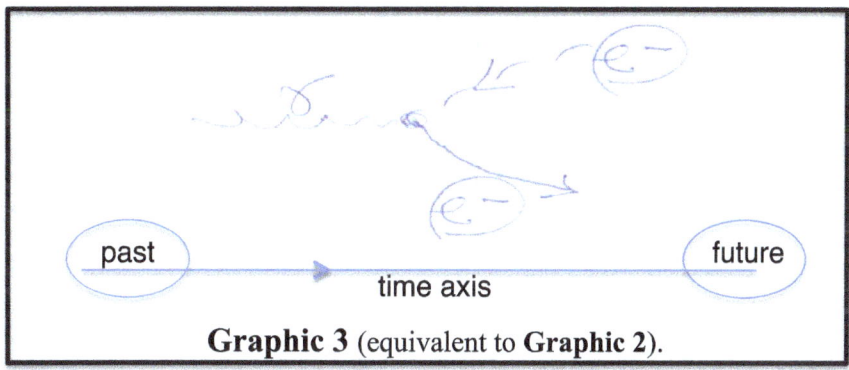

Graphic 3 (equivalent to **Graphic 2**).

The annihilation of an (e⁺e⁻) pair into a photon γ is illustrated in **Graphic 4**.

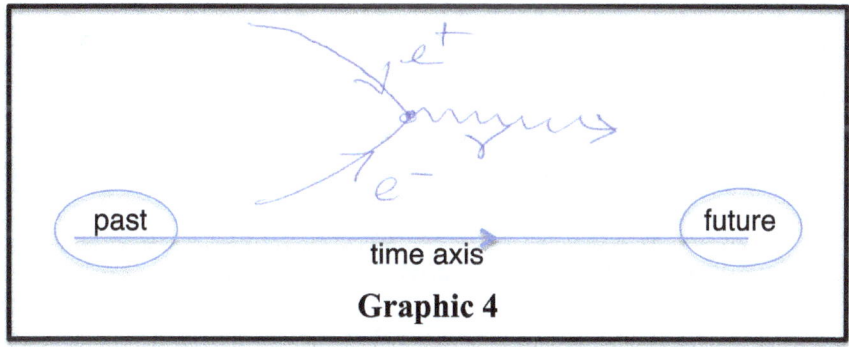

Graphic 4

The annihilation process of **Graphic 4** is equivalent to the sequence of events illustrated in **Graphic 5**. An electron is going from past to future and emits a photon. From this moment it becomes an electron (e⁻) going from future to past.

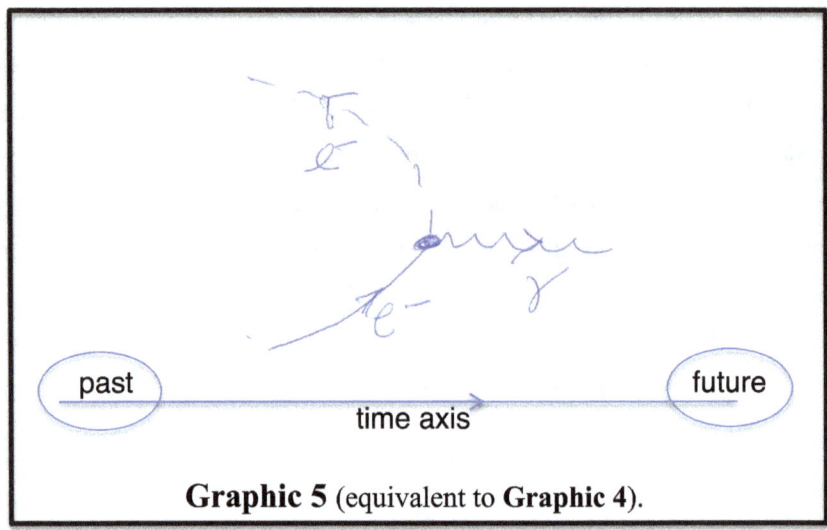

Graphic 5 (equivalent to **Graphic 4**).

Putting together **Graphics 2** and **4** we get **Graphic 6**.

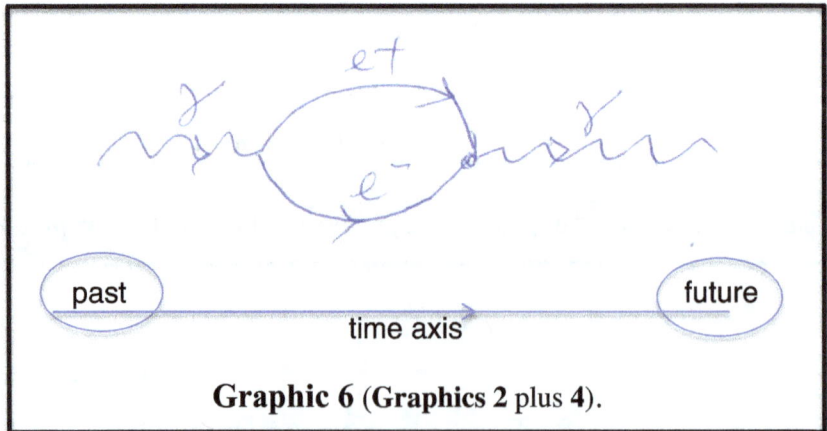

Graphic 6 (**Graphics 2** plus **4**).

Putting together **Graphics 3** and **5** we get **Graphic 7**.

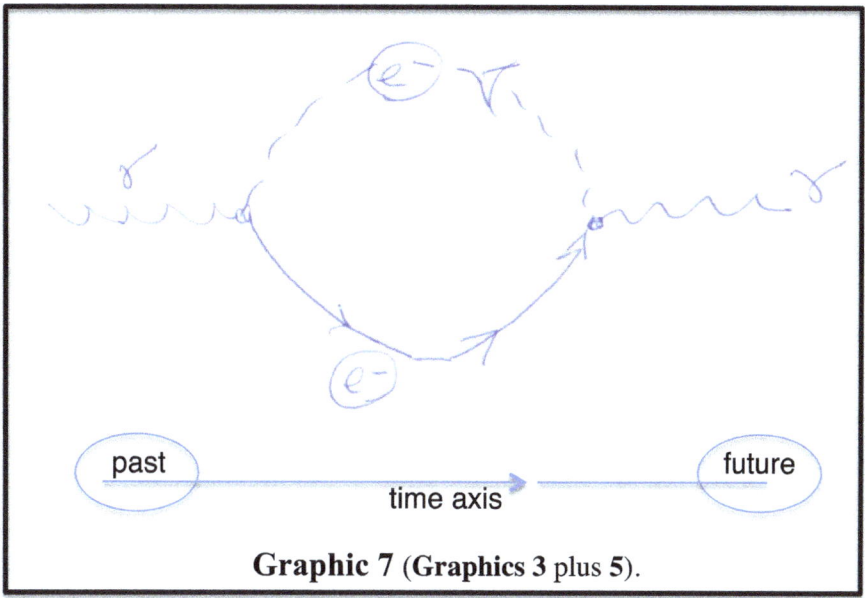

Graphic 7 (**Graphics 3** plus **5**).

This is equivalent also to **Graphic 8**.

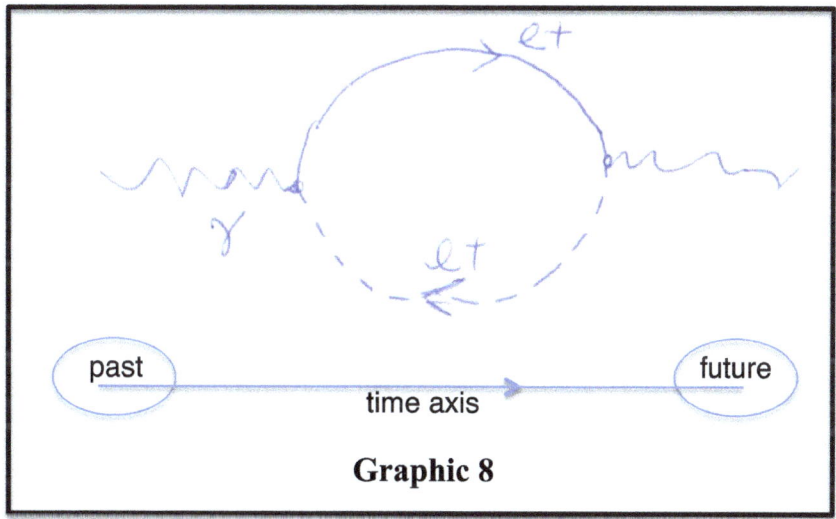

Graphic 8

The conclusion predicted by Dirac is that the existence of (e⁺e⁻) pair production from a γ–ray (**Graphic 2**) implies the existence of the (e⁺e⁻) annihilation into a γ–ray (**Graphic 4**).

Combining production and annihilation there is a "rotation" of charges (negative ≡ e⁻ and positive ≡ e⁺) which produces the "vacuum polarization" effect, illustrated in **Graphic 9**, first computed by Victor Weisskopf for its effects on the energy levels of hydrogen (**Graphic 1**).

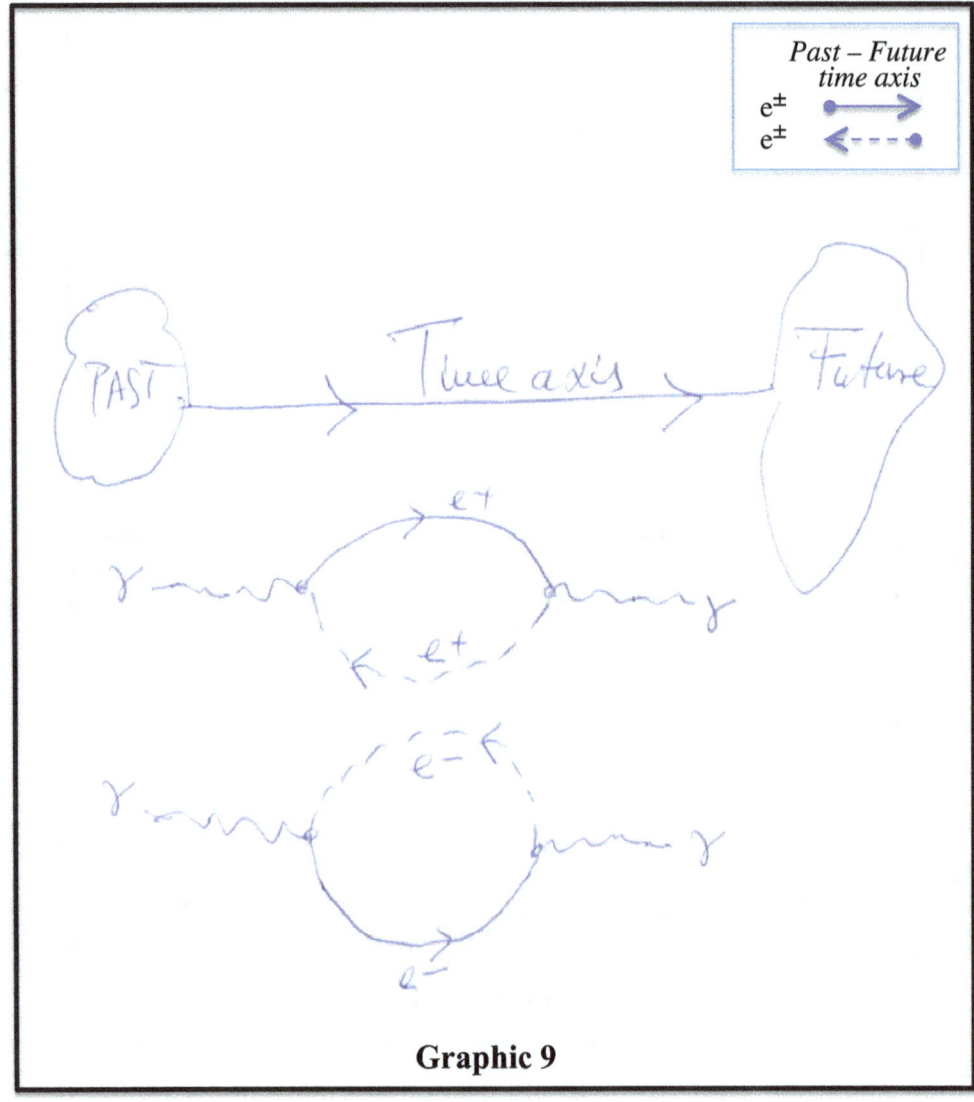

Graphic 9

Putting together particles going from past to future with particles going from future to past, the rotation of an electron or of an antielectron produces the polarization effect called "vacuum polarization".

When Weisskopf was working on this effect, many of his colleagues were of the opinion that he was working on something totally irrelevant for the future of physics.

At a much less difficult theoretical level – as mentioned before – no one was able to predict the virtual process (**Graphic 10**) experimentally discovered in 1947: an electron produces a photon and absorbs the same photon.

Graphic 10

This effect for the hydrogen energy levels (**Graphic 11**) was (and is) with a positive sign and more than ten times larger: **+ 1000 Mc/sec** than the "vacuum polarization".

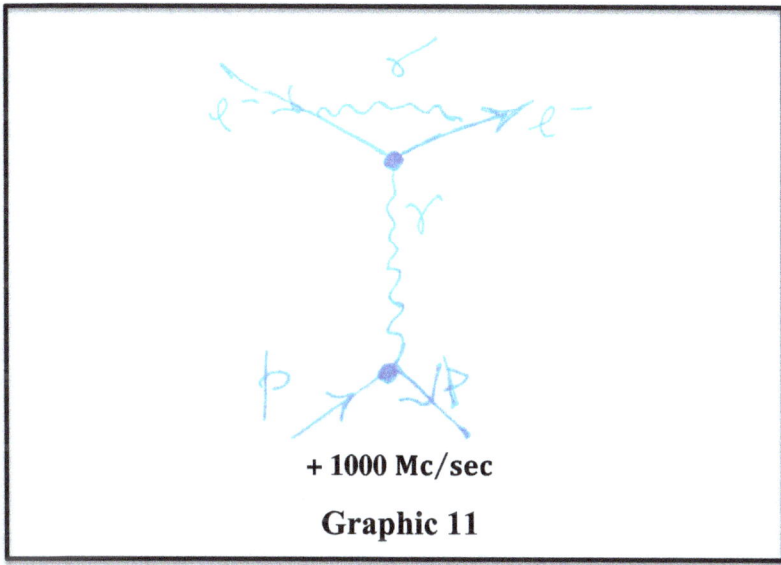
+ 1000 Mc/sec
Graphic 11

The experimental uncertainty was **± 100 Mc/sec**, a value larger than the one computed by Weisskopf for the "vacuum polarization" in hydrogen (**Graphic 1**).

No one could have imagined what is reported in **Figure 12**: the gauge unification and the GAP.

The present status of the Unification of the three gauge couplings, $\alpha_1\ \alpha_2\ \alpha_3$, is illustrated in **Figure 13**. The mathematics needed is given in **Figure 14**, which is the most accurate description of these physics processes. Using the "best" experimental values it comes out the existence of the GAP, reported in **Figure 15**. The GAP is between the Planck energy, E_{Planck}, and the two energy levels, E_{GUT}, where the three gauge couplings converge and E_{SU}, the energy where the RQST (Relativistic Quantum String Theory) puts the origin of the gravitational forces. The existence of this GAP is based on the use of all "best values", as discussed in the original paper [65].

The Gran Sasso Underground Laboratory [66] is for the study of cosmic events such as the Supernovae, the stability of matter and the neutrino mixing problems. All this physics has its roots at the extremely high energies: E_{GUT} and E_{Planck}.

These theoretical achievements are all the result of "virtual physics". This physics was considered useless in 1934. No one could have imagined that, thanks to the "virtual physics", the existence of the energy threshold of the Superworld would have been computed.

These new horizons were opened by the discoveries originating in the Blackett group (**Figures 6, 11, 12**).

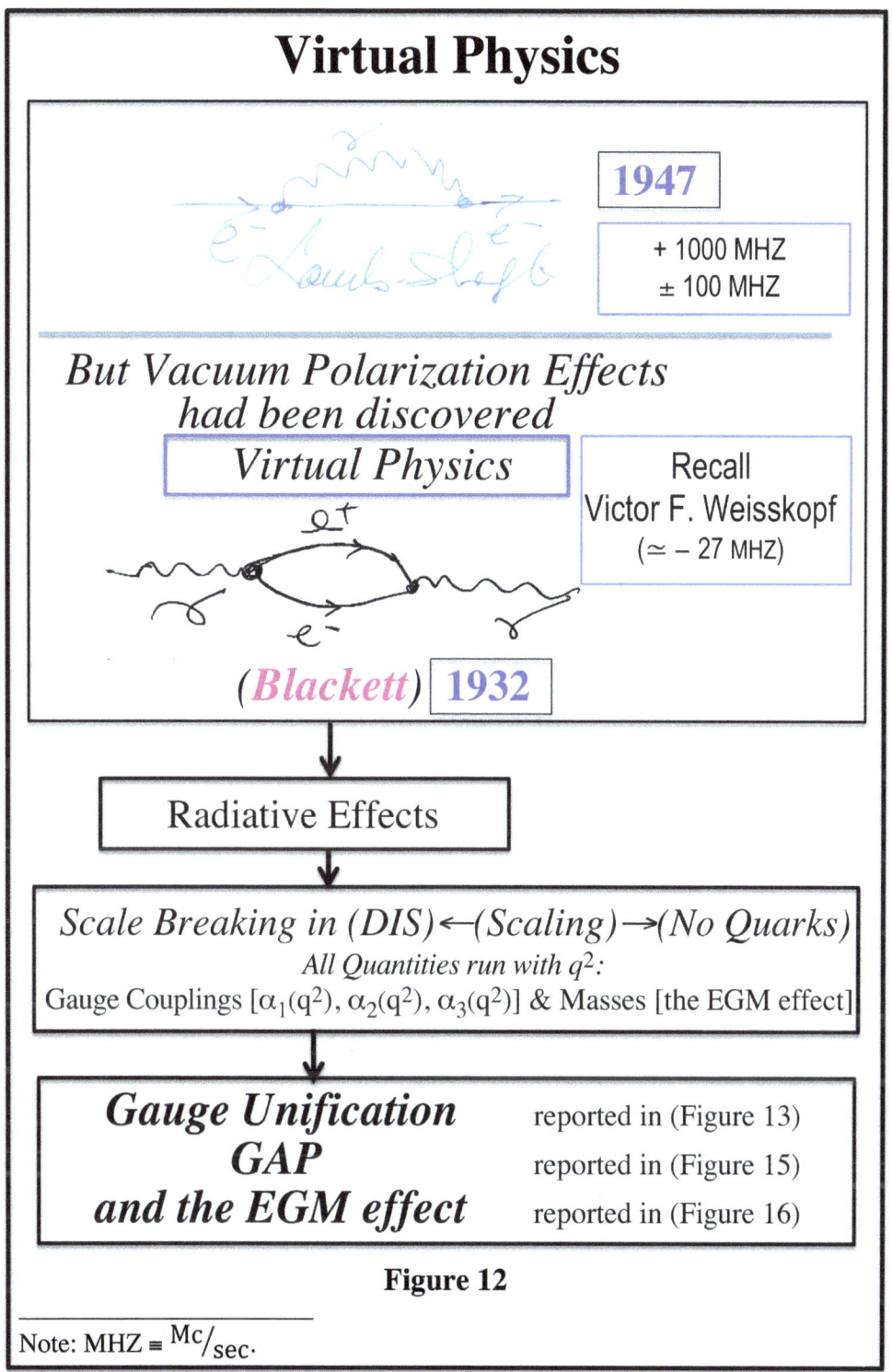

Figure 12

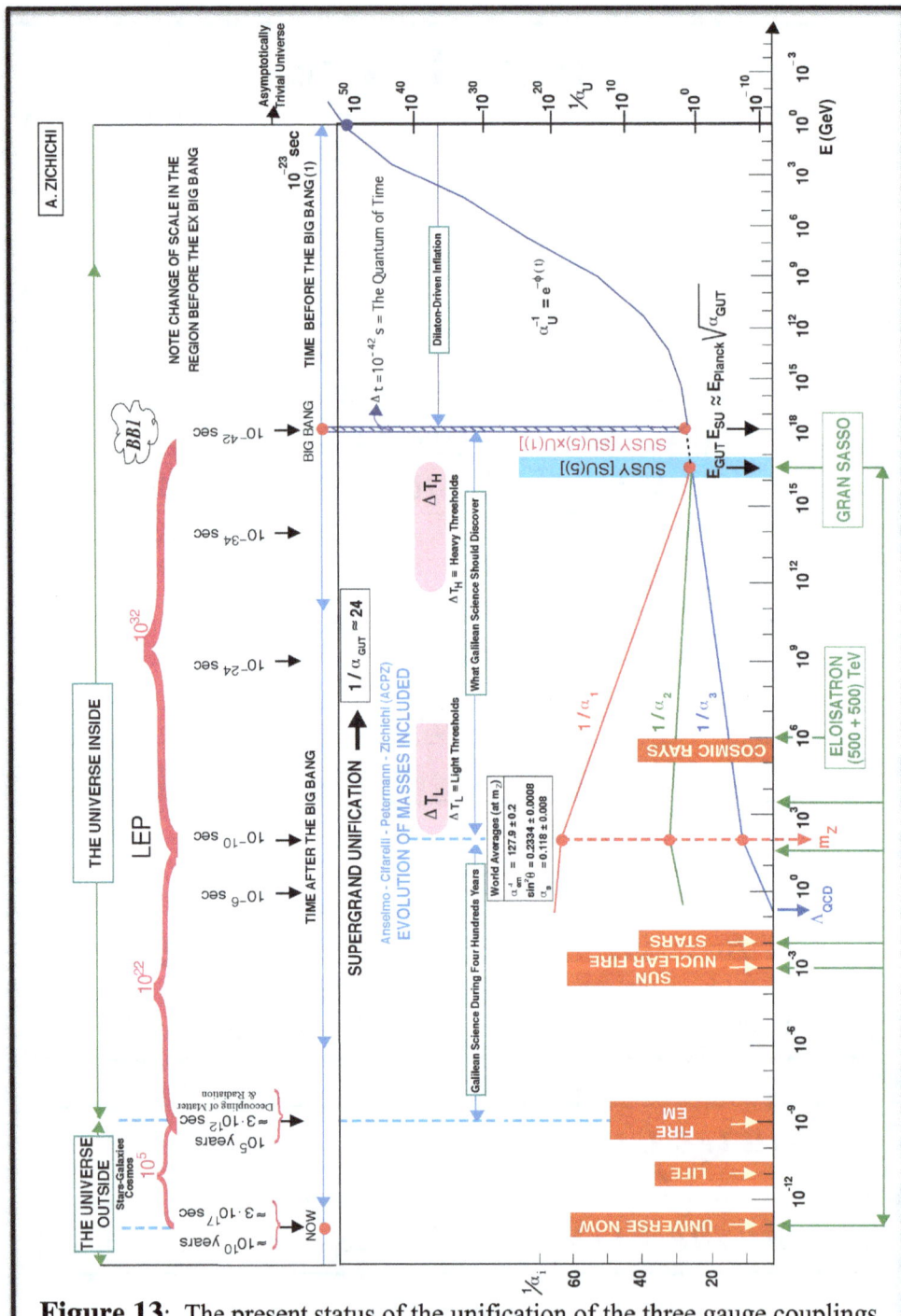

Figure 13: The present status of the unification of the three gauge couplings α_1 α_2 α_3 is reported. This is the most accurate description of the unification whose mathematics is in **Figure 14**.

THE UNIFICATION OF ALL FUNDAMENTAL FORCES AND THE EGM EFFECT

The three lines (α_1^{-1}, α_2^{-1}, α_3^{-1}) in **Figure 13** result from calculations executed with a supercomputer using the following system of equations:

$$\mu \frac{d\alpha_i}{d\mu} = \frac{b_i}{2\pi} \alpha_i^2 + \sum_j \frac{b_{ij}}{8\pi^2} \alpha_i \alpha_j$$

α_i, α_j (with i = 1, 2, 3; and J = 1, 2, 3 but i ≠ j).

This is a system of coupled non-linear differential equations where the existence of the Superworld is included. This system describes how the gauge couplings ($\alpha_1, \alpha_2, \alpha_3$) vary with "$\mu$", the basic parameter which depends on the energy of the elementary process, from the maximum level of Energy (Planck Scale) to the energy level of our world. **During more than ten years (from 1979 to 1991), no one had realized that the energy threshold for the existence of the Superworld was strongly dependent on the "running" of the masses**.

This is now called: **the EGM effect** (from the initials of Evolution of Gaugino Masses). This effect produces a factor 700 in the threshold for the lightest supersymmetry particle. No one knows the value of the threshold Energy for the production of the lightest supersymmetry particle. The EGM effect lowers this value by a factor 700. Suppose that somebody would say that the threshold is at 700 TeV, thanks to the EGM effect this value is going to be 1 TeV.

Figure 14

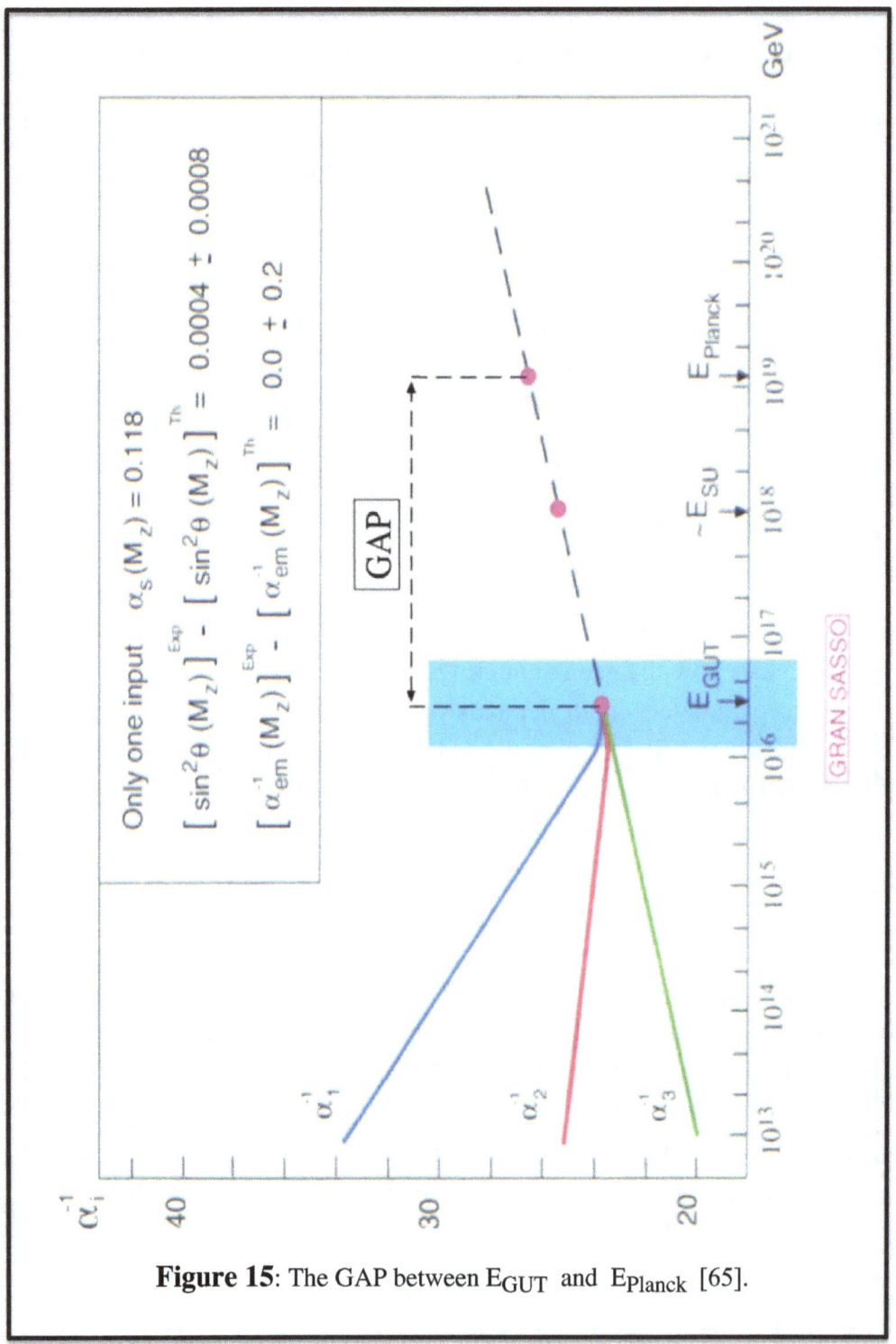

Figure 15: The GAP between E_{GUT} and E_{Planck} [65].

And that is not all. The predictions for the Supersymmetry threshold received a boost when the EGM (Evolution of Gaugino Masses) effect was discovered [67].

This effect lowers the threshold for the existence of the Superworld by a factor 700, as shown in **Figure 16**.

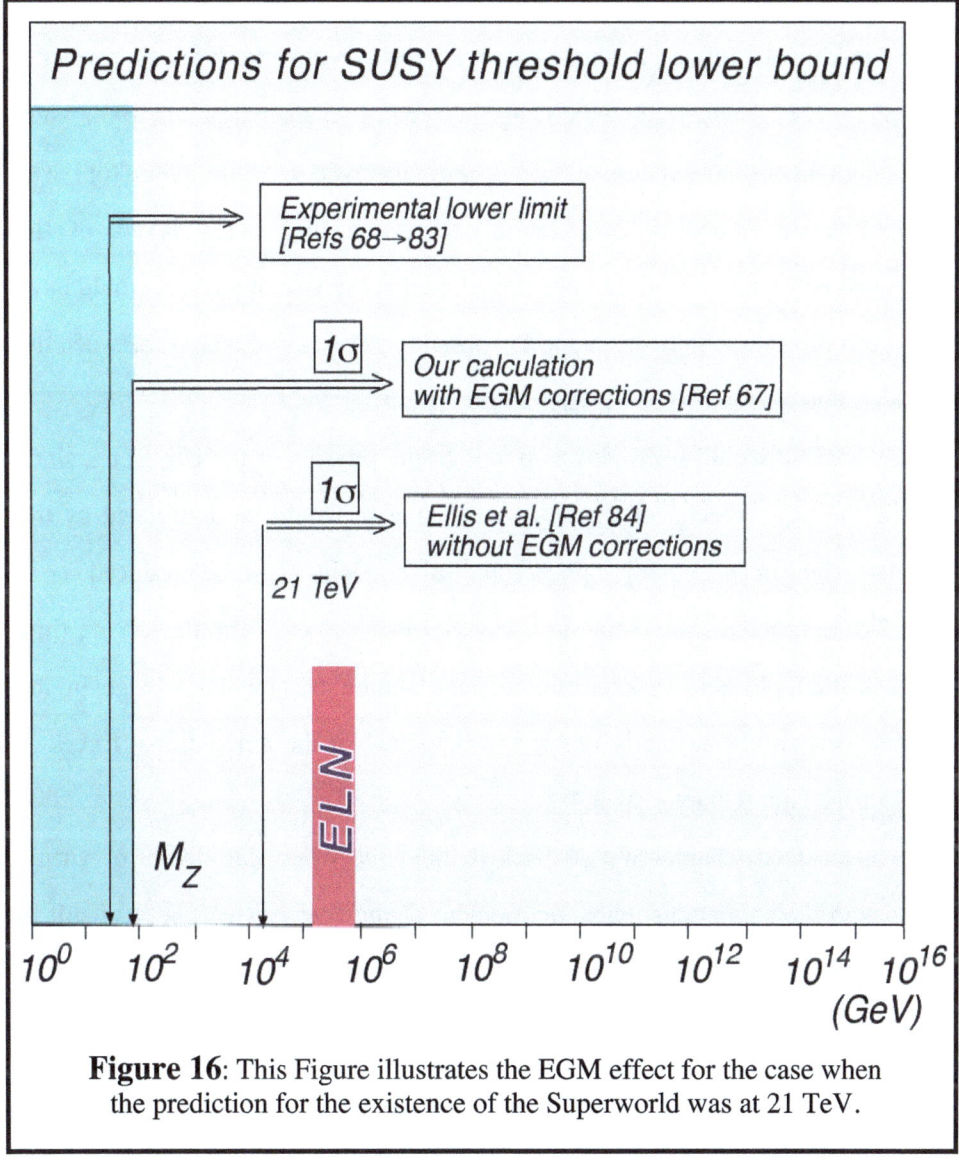

Figure 16: This Figure illustrates the EGM effect for the case when the prediction for the existence of the Superworld was at 21 TeV.

This effect is quite interesting for the LHC. In fact, if somebody would "predict" (ignoring the EGM effect) that the energy threshold for producing the lightest particle of the Superworld should be at 700 TeV, and therefore out of the energy range where LHC could discover the first supersymmetric particle, thanks to the EGM effect, the threshold would become 1 TeV, perfectly achievable with the LHC.

Looking at **Figure 12**, once again we have to realize how relevant was the "prediction" of Blackett quoted on page 46 (end of Chapter III-2). In 1932 when the simultaneous production of (e^+e^-) was discovered, while Weisskopf was computing the "vacuum polarization" effect on the energy levels of the hydrogen atom, no one would have imagined that the virtual physics would have produced the gauge unification and the possible existence of a GAP between the two extreme energy levels: E_{Planck} and E_{GUT}.

The GAP would open new horizons in the physics of the Big Bang, since during the time interval of the GAP the Universe would be dominated by the gravitational forces, and only "primordial" Black Holes could be produced.

These problems are very far away from the physics frontier of the time when virtual physics was starting to be a source of new knowledge in the Subnuclear world but totally ignored in the study of the gravitational forces as we will see later (Chapter III-3.4).

In the next Chapter we go back to the time when the effects of virtual physics in the anomalous magnetic moment of the "heavy electron" had still to be implemented.

III-3.1 – THE 1ˢᵗ DISCOVERY AT CERN IN THE PHYSICS OF HIGH PRECISION. CONSEQUENCES OF VIRTUAL PHYSICS ON THE "HEAVY ELECTRON" CALLED MUON

The first discovery at CERN in the physics of high-precision is linked to Professor Blackett, since the "anomalous" magnetic moment of the muon $(g-2)_\mu$ could not have been theoretically predicted, if virtual physics had not been discovered (in 1932) and also as the youngest fellow of the Blackett group at CERN was strongly engaged in the venture to understand what really was the heavy electron called muon.

The muon is 200 times heavier than the electron and, at that time, the origin of the mass was unknown (as it is today) but believed to be due to unknown interactions in addition to the well-known QED (Quantum ElectroDynamics) effects. For example, electromagnetism was believed to produce mass differences Δm of the order of MeV, such as the (pn) mass difference. What was the origin of the two orders of magnitude (10^2 MeV mass difference) between the muon ($\simeq 100$ MeV) and the electron ($\simeq 0,5$ MeV) masses?

Zel'dovich computed the virtual physics effects on the $(g-2)_\mu$, assuming for the muon the existence of interactions of reasonable strength in addition to the (EM) one. The effect of these interactions could have been quite large; of the order of a percent. After all, the muon was 200 times heavier than the electron. What needs to be done?

Measure the "anomalous" magnetic moment of the muon, $(g-2)_\mu$ (mentioned in Chapter III), with good accuracy. This accuracy had to be as high as possible: something of the order of few parts per million for the muon magnetic moment. There were different proposals: the screw magnet was advocated by Leon Lederman. My choice was the "flat magnet". This had already been tried by R. Nikitin in Dubna (see later) and failed because high precision magnetic fields (at the 10^{-4} accuracy level) were needed. These fields had to be of polynomial form in order to allow injection, storage and ejection in an adiabatic way: i.e. in such a way that the transition for example from one field (such as injection) to storage takes place as all other transitions in the softest possible way. The basic problem was the magnetic field for storage, which needed to be as long as possible: the muon momentum had to rotate thousands of times, since the mismatch between linear momentum $\vec{P}_\mu$ and magnetic moment $\vec{\mu}$

was of the order of per mill,

$$\left(\frac{\alpha}{2\pi}\right) \simeq 10^{-3}$$

per revolution.

A total of 7 polynomial magnetic fields were needed: one for injection, two for transitions from injection to storage, one for storage, two for transitions from storage to ejection and one for ejection. These seven magnetic fields had to be all at high precision levels. The theoretical problem of the polimomial magnetic fields needed had to be translated into the effective realization of these magnetic fields.

The experts knew only one technology in order to have the needed high precisions polinomial magnetic fields: high precision milling-machines to obtain the needed shape for the iron poles of a "flat magnet". The estimated time: many years (5–10) of very hard work. The estimated cost: extremely expensive.

Proceedings of the 1962 International Conference on High Energy Physics, pp. 476-480

(g-2) AND ITS CONSEQUENCES

G. Charpak, F. J. M. Farley, R. L. Garwin, T. Muller, J. C. Sens and A. Zichichi

CERN, Genève

(presented by A. Zichichi)

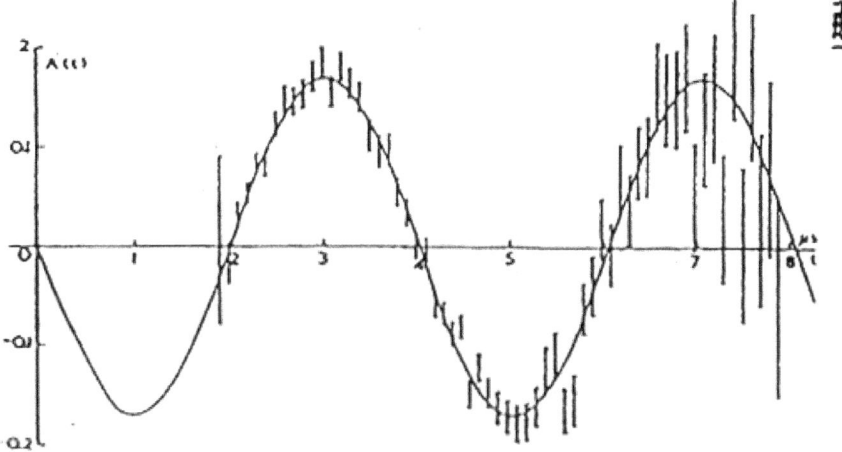

Fig. 5 Experimental data of the $(g-2)$ experiment. Observed electron decay asymmetry $A(t)$ as a function of storage time. The curve represents the best fit of the data.

RESULTS OF G - 2 EXPERIMENT

Experimental	Theoretical
	$\frac{g-2}{2} = \frac{\alpha}{2\pi} + 0.75 \left(\frac{\alpha}{\pi}\right)^2$
$\frac{g-2}{2} = 0.001162 \pm 0.000005$	$= 0.001161 + 0.000004$ $= 0.001165$
Muon mass $= (206.768 \pm 0.003)m_e$	
Charge of muon $= (1.00000 \pm 0.00005)e$	18 December 1961.
Charge of ν_μ $= (0.00000 \pm 0.00005)e$.	

Figure 17: The first high precision measurement of QED virtual physics effects outside the electron and photon world.

In order to overcome this difficulty it was necessary to invent a new technology 100 times faster for construction and 100 times less expensive, as reported in Chapter III-3.2. The experimental results are reported in **Figure 17** and represent the first high precision measurement of QED (Quantum ElectroDynamics) virtual physics effects in a domain outside the standard QED phenomena based only on electrons and photons [52].

As mentioned in Chapter III-2, all high precision measurements of the muon's electromagnetic properties are in the References [49–53].

These high precision experimental data on the anomalous magnetic moment of the muon could be obtained thanks to the invention which will be described in the next chapter.

III-3.2 – THE 1ˢᵗ INVENTION AT CERN FOR THE PRODUCTION OF HIGH PRECISION MAGNETIC FIELDS THAT IS 10^2 TIMES FASTER FOR CONSTRUCTION AND 10^2 TIMES CHEAPER THAN ALL EXISTING TECHNOLOGIES

And now the invention of a new technology for the experiment on the muon electromagnetic properties.

The former youngest member of the Blackett group invented the "shimming" technology: use very thin (microns in width) mu-metal sheets to achieve the "high accuracy" in the shape of the "flat magnet" poles. Where to get these extremely thin mu-metal sheets? They were familiar to me: I was using them to protect the photomultipliers from the secondary residual magnetic fields of our magnets. This is the origin of the "new technology" [85] which allowed (and still allows) the construction of high precision magnetic fields 10^2 times cheaper and 10^2 faster to construct than the then-known technology.

The schematic drawing of the largest and highest precision "flat" magnet of the world is reported in **Figure 18**. A photo of the magnet is in **Figure 19**.

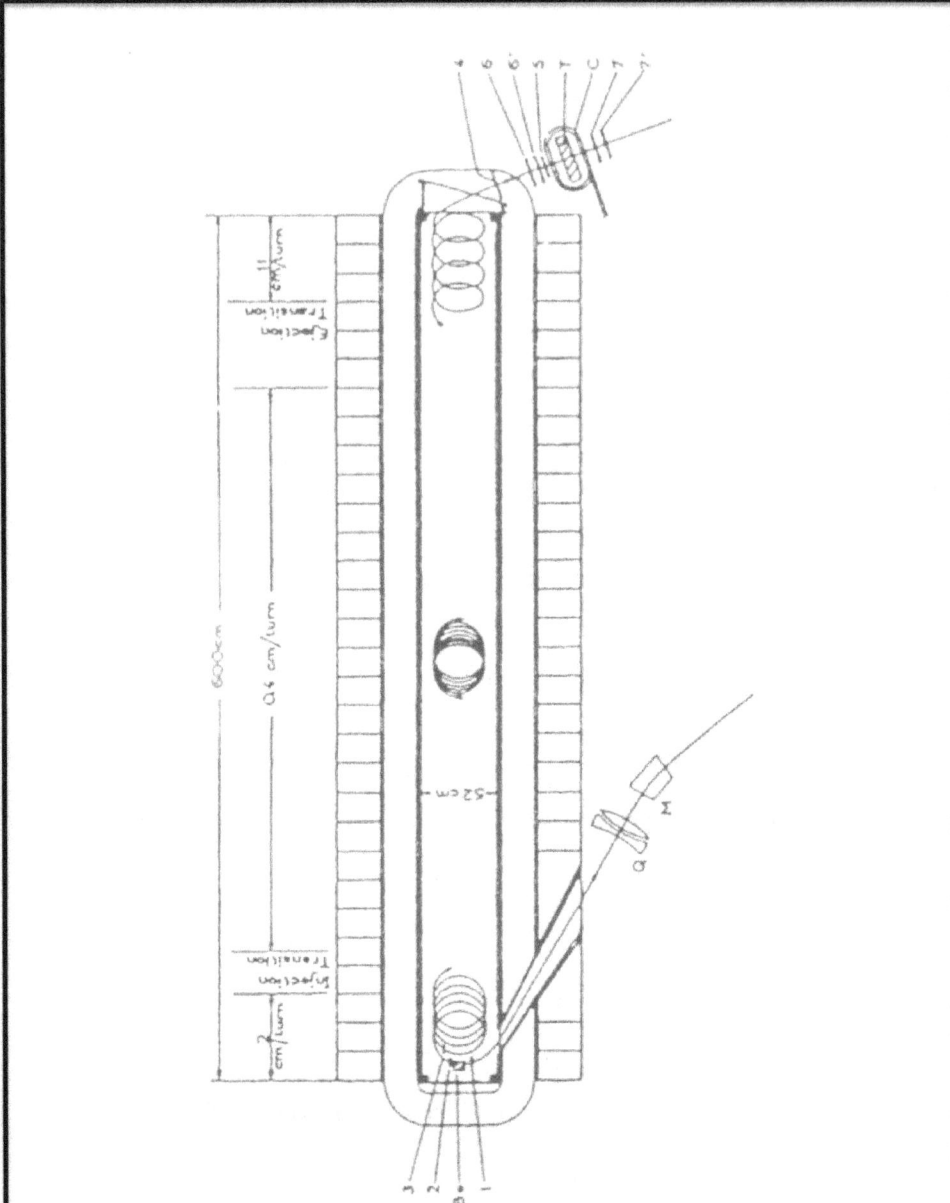

Figure 18: General plan of the 6-metre magnet. *M*: bending magnet; *Q*: pair of quadrupoles; 1, Be, 2, 3: injection assembly consisting of Be-moderator and counters 1, 2, 3; *T*: methylene-iodide target; counters 66', 77': "backward" and "forward" electron telescopes. A stored and ejected muon is registered as a coincidence 4, 5, 66' $\overline{7}$, gated by a 1, 2, 3 and by either a forward or backward electron signal.

Figure 19: A photo of the six-metre "flat-magnet" where a sequence of high precision magnetic fields has been implemented using what Feynman liked to call the "trick" of the "shimming technology".

This 6-metre magnet allowed the first high precision measurement of the electromagnetic properties of the muon, i.e. its anomalous magnetic moment (**Figure 17**).

The muon (g–2) gave me the opportunity to meet R.P. Feynman.

Dick Feynman was a theorist very interested in what he liked to call experimental "tricks". He was very much pleased with the idea of constructing high precision polynomial magnetic fields using the "trick" which avoided very expensive and time-consuming highly accurate machining.

As mentioned before, Nikitin came to CERN and visited the (g–2) flat magnet, not the 6-metre long one, but the "Liverpool" magnet, about 1-metre long. This magnet was graciously lent by Liverpool University to CERN in order to allow our tests of the "shimming" technology. Nikitin told me that he had tried to use a flat magnet for magnetic "storage" and failed, thus giving me the friendly advice to stop wasting time.

On the contrary, Feynman, in Weisskopf's office, was encouraging me: "*Never follow the advice of people who have failed in what you want to do.*"

Weisskopf had told him that I was the youngest fellow of the Blackett group. Feynman knew the Blackett effect in the Mediterranean battle (see Chapter V) and admired the physics discoveries of the Blackett group. He knew that Blackett was teaching his fellows to be proud of their wok.

This is how Feynman became my friend: Feynman came to the Ettore Majorana Subnuclear Physics School during the hard times when the school was holding its second course (1964). At the 1964 School he said: "*So in both theory and experiment, to improve the physics of the future, I would like to see a little more pride in the work*" [86].

III-3.3 – THE 1ˢᵗ PROOF THAT CERN COULD COMPETE WITH WELL ESTABLISHED FAMOUS LABS AND WIN

The proof that the new European Institution, called CERN, was able to perform experiments in fields where other well established Laboratories had failed is in the measurement of the "weak" charge. The problem was to establish if the particles carrying the strong charge (hadronic) with and without the new charge called "strangeness" were carrying the correct fraction of the "weak" charge. The high precision measurement of the 100% weak coupling was needed without experimental uncertainties.

These were times when the weak couplings of the "strange particles" was found to be different from the non-strange weak coupling. Was it really true that the "weak" charge was split into many pieces? Different fractions being given to different particles? The full weak coupling had (and still has) as a key experiment the muon lifetime [54]. The experimental results obtained at Carnegie and at Chicago were in great difficulty because of rate dependent effects which invalidated the data.

The experimental results obtained at CERN are shown in **Figure 20** [54]. The data show that the rate dependent trouble was overcome.

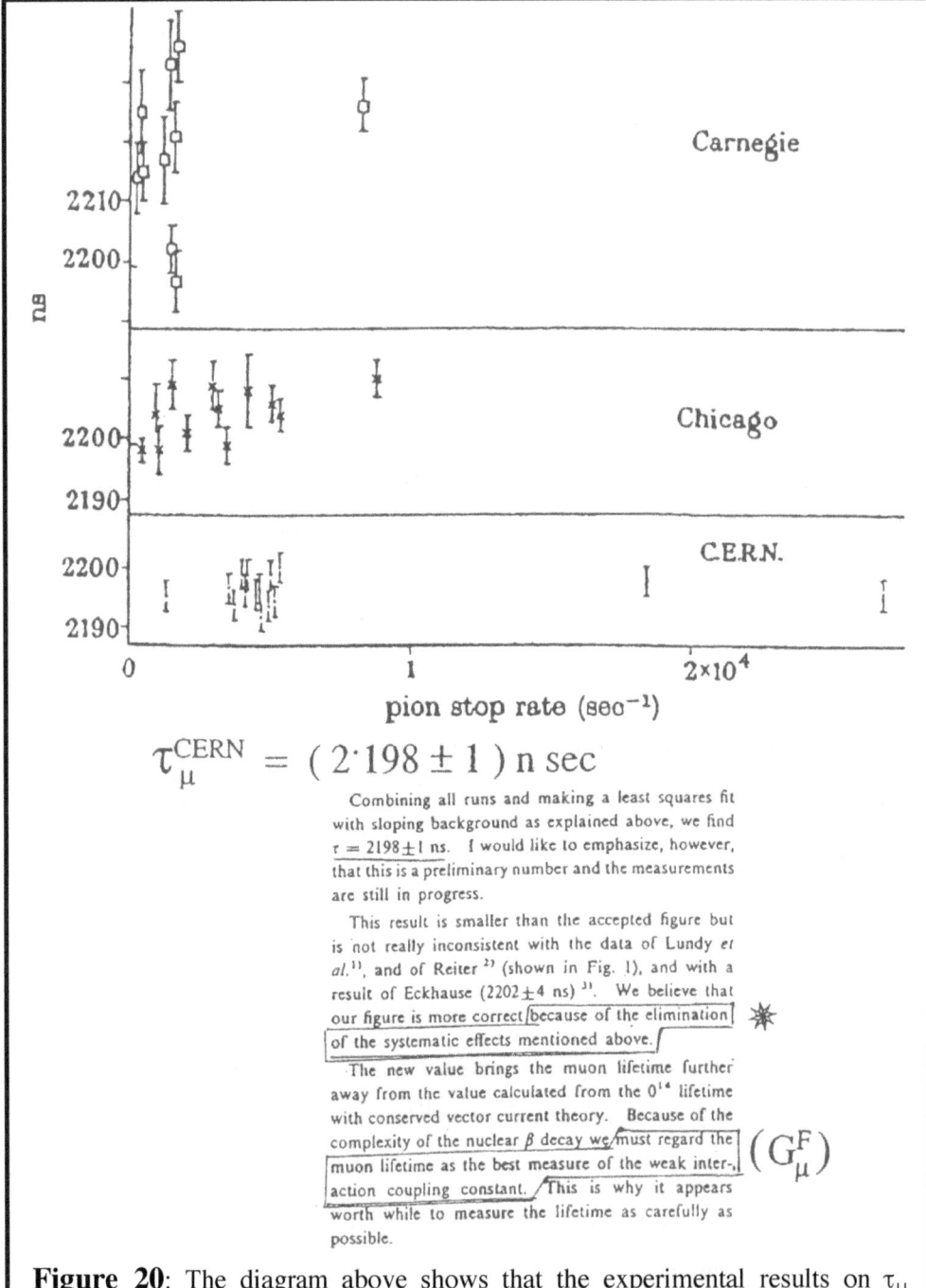

$$\tau_\mu^{CERN} = (2\cdot198 \pm 1)\text{ n sec}$$

Combining all runs and making a least squares fit with sloping background as explained above, we find $\tau = 2198 \pm 1$ ns. I would like to emphasize, however, that this is a preliminary number and the measurements are still in progress.

This result is smaller than the accepted figure but is not really inconsistent with the data of Lundy et al.[1], and of Reiter[2] (shown in Fig. 1), and with a result of Eckhause (2202 ± 4 ns)[3]. We believe that our figure is more correct because of the elimination of the systematic effects mentioned above.

The new value brings the muon lifetime further away from the value calculated from the 0^{14} lifetime with conserved vector current theory. Because of the complexity of the nuclear β decay we must regard the muon lifetime as the best measure of the weak interaction coupling constant. (G_μ^F) This is why it appears worth while to measure the lifetime as carefully as possible.

Figure 20: The diagram above shows that the experimental results on τ_μ obtained in Chicago and Carnegie were affected by a rate dependent systematic effect which invalidates the data. The CERN result is the first without this trouble.

And this is not all. The measurement of the e^+ polarization in μ–decay gave a disagreement with the standard weak properties of the muon. This disagreement was proved to be incorrect [55].

The "perfect" electromagnetic and weak properties of the muon were discussed during many years with a great friend of mine Abdus Salam who – as said in Chapter I, page 10 – joined the Blackett group after me. Here is what Salam said in his 1979 Nobel Lecture [*Gauge Unification of Fundamental Forces*, Abdus Salam, *Nobel Lecture*, 8 December 1979, from *Nobel Lectures, Physics 1971-1980*, Stig Lundqvist (ed), World Scientific (1992)]: «*Zichichi had been badgering me since 1958 with persistent questioning of what theoretical avail his precise measurements on (g–2) for the muon as well as those of the muon lifetime were, when not only the magnitude of the electromagnetic corrections to weak decays was uncertain, but also conversely the effect of non-renormalizable weak interactions on "renormalized" electromagnetism was so unclear*».

Having proved that the muon behaved (and still now behaves) like a "perfect" QED particle with a "perfect" weak coupling, the former youngest fellow of Professor Blackett started thinking what to do next. The next step was the search for the third lepton [56], discussed in Chapter II-3 with reference to the ethics in our Science.

Figure 21: The set-up able to detect simultaneously (e^+e^-), ($\mu^+\mu^-$) and ($\mu^\pm e^\mp$) final states from ($\bar{p}p$) annihilation.

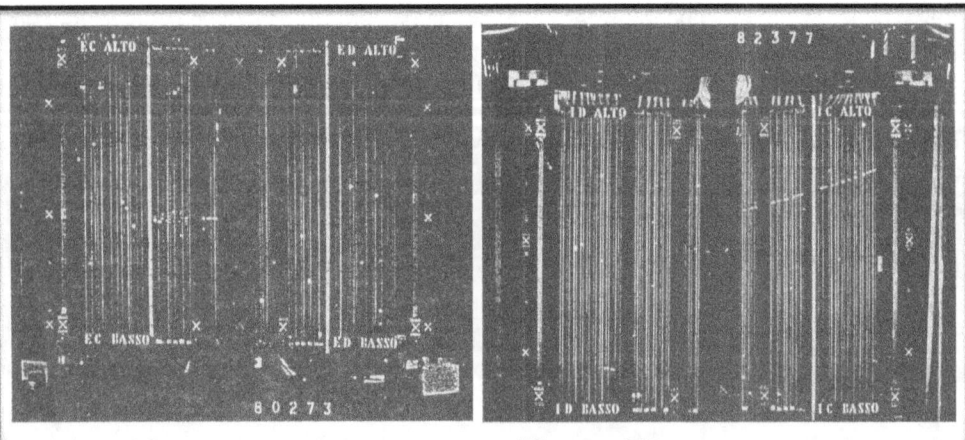

Figure 22: Typical electron event. **Figure 23:** Typical muon event.

These pictures are from the CERN-Bologna-Frascati set-up where the ADONE collider was the source of the ($\mu^\pm e^\mp$) final states.

III-3.4 – THE THIRD LEPTON AND THE PROBLEM OF GÖDEL IN PHYSICS

The third lepton has a fundamental role concerning the existence of the third Family of fundamental particles (quarks and leptons). If a third Family is there in the leptonic structure, it must be there also in the quark structure. The origin of the third Family is therefore in the search for the third lepton at CERN. The third lepton focused the attention of the physics community on the fact that, if a third lepton as heavy as the proton was there, no-one would have had seen it. In order to look for such a lepton it was necessary a beam of antiprotons at the highest possible energy and intensity [87]. The final state had to be a pair of acoplanar electron and muon. Many years of work, were needed and new technologies had to be invented. This is how the "preshower" technology for the electron detection was invented [88], the muon "punch-through" technology was studied [89], the first lead scintillator telescopes (now called calorimeters) were designed, built, and studied in detail [90] and the most precise TOF (Time-Of-Flight) detector technology was invented, designed and built.

In the **Figure 21** of page 72 there is a photo of the CERN set-up where $(e^{\pm}\mu^{\mp})$ pairs could be detected. The third lepton was called by the Bologna-CERN group $HL^{\pm}$ and was assumed to carry a new lepton quantum number, together with its neutrino, ν_{HL}. As a result, it would decay via the two possible reactions (one with $e^\pm$, the other with $\mu^\pm$)

$$\left\{ \begin{array}{l} HL^+ \to e^+ \, \nu_e \, \bar{\nu}_{HL} \; ; \quad HL^+ \to \mu^+ \, \nu_\mu \, \bar{\nu}_{HL} \\ HL^- \to e^- \, \bar{\nu}_e \, \nu_{HL} \; ; \quad HL^- \to \mu^- \, \bar{\nu}_\mu \, \nu_{HL} \, . \end{array} \right\}$$

The best way of searching for pair production of HL was therefore to look for events with nothing but acoplanar pairs of electron and muon observed in the final state plus missing momentum. The proposed detector was based on the techniques able to select electrons (e) and muons (μ) with powerful technologies against the "background" of other particles. The selection power of the Bologna-CERN group was at the level of 10^{-4}: one part in ten thousands parts (10^{-4}), a world record. These technologies were developed by the Bologna-CERN group many years before the proposal by Kobayashi and Maskawa (KM) [19d].

The third lepton was being searched for at CERN and at Frascati. The search at CERN (**Figure 21**) was using proton-antiproton annihilation [19a] in the early sixties, a decade before KM. At Frascati [19b] (see **Figure 24** and **22, 23**) six years before KM.

20 March 1967: The INFN-Bologna Proposal to Search for Heavy Leptons

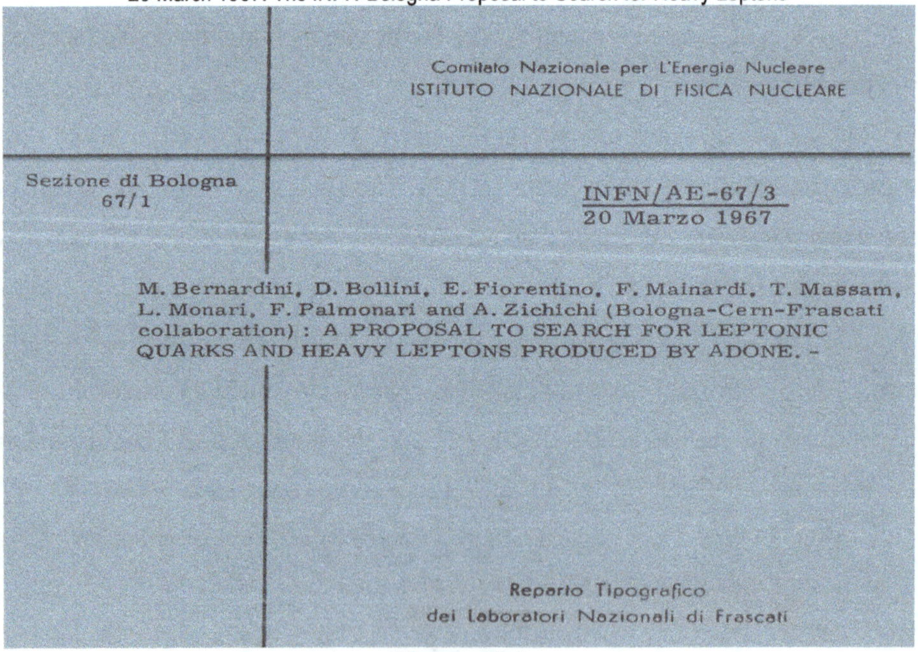

Comitato Nazionale per L'Energia Nucleare
ISTITUTO NAZIONALE DI FISICA NUCLEARE

Sezione di Bologna 67/1

INFN/AE-67/3
20 Marzo 1967

M. Bernardini, D. Bollini, E. Fiorentino, F. Mainardi, T. Massam, L. Monari, F. Palmonari and A. Zichichi (Bologna-Cern-Frascati collaboration) : A PROPOSAL TO SEARCH FOR LEPTONIC QUARKS AND HEAVY LEPTONS PRODUCED BY ADONE. -

Reparto Tipografico
dei Laboratori Nazionali di Frascati

Figure 24

And this is not all. The KM proposal was to look for a new quark of the third Family. The proposal by the Bologna-CERN was to search for a new lepton with its own neutrino. If the third lepton mass was in the one billion electron-Volts (1 GeV) it would have been discovered at Frascati in 1970 [19c] (see Figure below).

The most favourable mechanism for the production of a heavy lepton HL is

(1) $\qquad e^+e^- \to HL + \overline{HL}$,

which, in the one-photon approximation, is described by the Feynman diagram

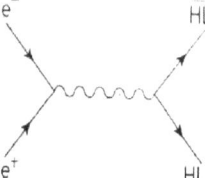

By analogy with muon decay, we may expect the predominant decay channel to be

(2) $\qquad HL^\pm \to e^\pm \nu_e \nu_H$,

$\qquad\qquad \to \mu^\pm \nu_\mu \nu_H$

where ν_H indicates the neutrino proper to the heavy lepton HL. If this is the case, and the two modes are, as expected, equally probable, then there is $\sim 50\%$ probability that the final state from reaction (1) consists of a noncollinear $e^\pm \mu^\mp$ pair.

From: "*Limits on the Electromagnetic Production of Heavy Leptons*", Lettere al Nuovo Cimento 4, 1156 (1970).

Figure 25: From [19c].

The search for the heavy lepton at CERN using the annihilation reaction of protons against antiprotons gave the discovery of the existence of the electromagnetic "time-like" structure of the proton [19a]. If the proton was an "elementary" particle (made of nothing but of itself) this "structure" shouldn't be there. This lack of structure was given for granted in 1947, when all known elementary particles with "strong" charges were only three: proton, neutron and pion (the Yukawa's meson) as reported in Chapter III-2 (see *"The Yukawa's meson: the first example of the nuclear glue"*, Appendix 10). No one could imagine what we know now: the proton and the neutron have in their intrinsic structure the Subnuclear Universe.

The discovery of the "time-like" structure of the proton gave to the Bologna-CERN group the message that the proton-antiproton annihilation could not be the best way in order to produce the third lepton. The best annihilation had to be between two effective "elementary particles". The only elementary particle in the market was the electron. The annihilation of an electron with its antiparticle (the antielectron) was going to be the best source for the production of the third lepton. This is why the Bologna-CERN group moved to Frascati where a new collider (ADONE) was going to be implemented for electron-antielectron annihilation.

No-one knew what the mass of this new heavy lepton had to be. Still now no-physicist is able to explain why the third lepton mass is 1,7 GeV. The Bologna-CERN group was hoping that the mass of the third lepton was at most, one billions electron-Volts, as heavy as the nucleon (proton and neutron), the heaviest particle known in the physics of strong interactions.

If this was the case the third lepton (see **Figure 26**) would have been discovered at Frascati in 1970, **three years before** the proposal by KM for the existence of a third Family of quarks and **five years before** the discovery of the third lepton at **SLAC** [20].

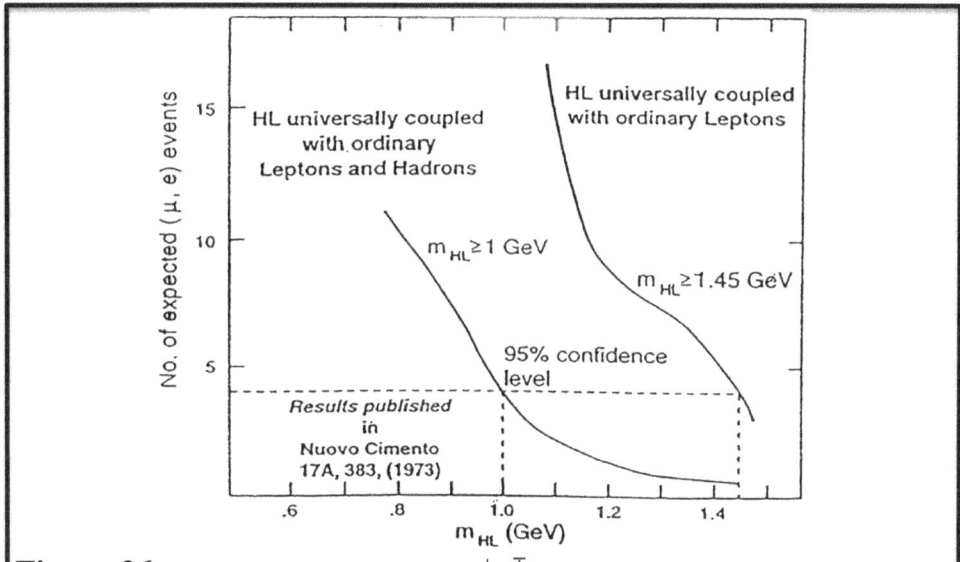

Figure 26: The expected number of $(e^{\pm}\mu^{\mp})$ pairs vs. m_{HL}, i.e. the heavy lepton mass, for two types of universal weak couplings of the heavy lepton [19e].

We have reported this sequence of events, as example of how in our Science what you think it is right (for example that a third Family of fundamental particles must exist) can be demonstrated to be effectively right, or wrong. The answer is given only by the result of an experiment. In all we have said so far about the third lepton the problem was to understand the structure of the elementary constituents of the Universe. The Universe that Professor Blackett was discussing with his friend Bertrand Russell in terms of its fundamental structures.

We will see in the next chapter that, it is thanks to Russell that Gödel was introduced in our discussions.

In mathematical logic Gödel discovered that it is not correct to say that a statement can only be either right or wrong. There is a third possibility: that it is impossible to decide. In our Science, Physics, it is the experimental result which establishes the validity of a statement.

The most famous example refers to the existence of the smallest amount of **action**. If somebody claims that the smallest amount of action can be as small as zero, this statement is proved to be wrong by performing an experiment.

To establish if the third Family of fundamental particles is or is not in the structure of the Universe only the experimental result can say. In the discussions with Russell the problem was not for the existence of the third lepton but for the smallest amount of action, which cannot be smaller than the value experimentally established and now called Planck's constant.

The amount of action needed to bring during one millisecond one milligram of mass up (against the gravitational attraction) by one millesim of a millimetre is equal to ten billions of billions times the Planck's constant. A quantity which we cannot observe using our five senses.

We need our instruments which give tremendous more power to our five senses. And here we meet the problem which afflicted Mach who died convinced that Atomic Physics was not Science (see "*Mach died convinced that Atomic Physics was not Science*", Appendix 1). The reason being that we are not able to see the atoms (see "*Why we cannot see an atom*", Appendix 12). Thanks to our technological inventions the power of our senses is tremendously increased. The problem is that in order to establish the validity of a "sentence" we need our technological inventions.

We will discuss these problems. For the time being let us go back to Russell. The decision-making power, clearly distinguishes the Rigorous **Theoretical** Logic (Math) and the Rigorous **Experimental** Logic (Science). The consequence of this fundamental distinction was, for a young fellow like me, extremely interesting. I will report on different occasions (Chapters IV, XII and "*The Vienna Circle and Gödel's discovery*", Appendix 13) the Gödel discovery in order to call as much as possible the attention of the reader on this problem, that was so much focused by Blackett and Russell.

I hope I will be able to report at least the essence of these discussions since even now, many decades later, this is a fascinating problem far from being fully understood. In fact the discovery in 1916 of the Schwarzschild singularity [91] gave a totally unexpected novelty defined in 1957 "Black Holes" by John Wheeler [92]. He is the physicist who has been engaged on these problems with his collaborators during many decades [93–96]. The physics of the Black Holes does not allow Science to perform experiments in order to establish the validity of our Logic [97] unless new horizons can be open in the Physics of the Black Holes [98].

IV – BLACKETT AND RUSSELL (GALILEI, EINSTEIN, GÖDEL)

Thanks to **Lord Blackett** I had the privilege of spending an evening with **Bertrand Russell**, getting to know his views on us physicists engaged at the frontiers of human knowledge in order to understand the Logic of Nature.

Blackett invited me at his place along with his friend, Bertrand Russell, who at some point in the discussion said: "*You physicists, just to do an experiment, you would sell your soul to the devil.*" We will see later how Russell arrived at this extreme conclusion. In this chapter the quotations should be taken as conceptually correct. The exact wording of course not.

The meeting at Blackett's house was actually to discuss **Relativity** and the role played by **Galilei**. Blackett was not happy with what Russell had written in his book, the *ABC of Relativity*. Galilei was only cited once despite the fact that Galilei's Principle of Relativity was so well formulated that it included Electromagnetic Phenomena (unknown to Galilei). I was there because, from childhood, I had read all of Galilei's writings. How Professor Blackett knew this detail about my childhood I do not know. Probably it was given to him by Peter Astbury, a physicist of the Blackett group with whom I had been working at

Jungfraujoch and with whom I developed a close friendship. Once he visited my family in Trapani.

From the Galilei-Einstein theme, the conversation turned to a subject that was **gnawing at Russell's** mind: *Physicists, that vile cursed race of beings.* He did not say it out loud, but he certainly harboured the thought. At one point he even exclaimed: "*The day there will be a slowdown in the race for armaments you physicists will be in trouble. We do not know for how long this will continue. Therefore, your future is secure. However, if I could think of a world without any more conflicts between great powers, if I were you I would change job. I don't know when the* **Cold War** *between the two super powers will become "hot". We should do all we can to avoid this but you should know that, in a planet without the two opposing super powers, the race for armaments will subside and so will the financing of frontier physics research.*

The **political powers are not financing your projects to better understand the Logic of Nature. This belongs to us and the truth will always be in our hands. You physicists are getting all that attention from governments because of the potential applications of your research to war technologies.** *The day that interest will wane, there will be no more financing for your projects.*"

The way Russell came to say what he really thought of us, the physicists, could be of interest. The key point of the discussion was the role played by Galilei in the Principle of Relativity. As said before, the principle – as written by Galilei – included all possible experiments: none of them would have allowed anyone to know if he was in the port of Livorno at zero speed, or if the boat was moving at constant speed in open sea. Galilei wrote: "*The water will fall in a glass in the same way. The smell of incense is the same even if the ship goes at high speed, but steady.*" We know now that the water falling in the glass is hydrodynamics but the smell of incense is an electromagnetic phenomenon. This is why the Principle of Relativity formulated by Galilei includes – as mentioned before – electromagnetism. In order to discover the Maxwell equations (1873) and the fact that our five senses are effects due to the electromagnetic forces more than two centuries were needed.

At some point the discussion became very specialized and Russell said: "*Einstein told me this.*" And Professor Blackett: "*But the author is you.*" He turned to me saying: "*We had in hand the radar technique; it was up to us to discover the Lamb-shift. Pity you were not in my group.*" The "pity" was connected with the discovery of the strange particles' associated production discussed in Chapter II. The word radar prompted Russell to say. "*Yes I know that you want to convince this young fellow that you invented the radar technology to defend our country from Hitler. You physicists believe that the governments support your researches because you are the highest intelligence capable of discovering the truth. Well no. The reason why you are so successful in getting a lot of financial support is because your research work gives rise to more and more sophisticated war technologies. You physicists, just to do an experiment, you would sell your soul to the devil.*" Turning to me: "*What about you, my young fellow, what do you think of it?*" And looking at me, he added, "*Have you ever thought why my friend has succeeded in convincing the British government to finance the new laboratory in Geneva? You think that it is because you are searching for the truth. Listen please, you are young. If one day the East-West confrontation ends, you had better change activity. The governments will stop supporting your work.*"

Coming back to Russell's rather shocking point of view, Blackett told me later "*You know, Russell is mad at us because, after all,* **Kurt Gödel** *was a physicist, one of us.*" (see "*Wigner, von Neumann and Gödel*", Appendix 14). As everybody knows it is the famous Gödel theorem that destroyed, in 1931, the *Principia Mathematica* of Bertrand Russell and Alfred North Whitehead. Gödel's theorem (see "*The Vienna Circle and Gödel's discovery*", Appendix 13) fascinated me when I was trying to choose physics or mathematics for my future.

The discovery of Gödel turns the work of Russell and Whitehead into a monument to the greatest mathematical illusion of all times. In fact, the theorem of Gödel shows that any Axiomatic System representable by arithmetic either is <u>coherent</u> but incomplete, or else, if <u>complete</u>, cannot be coherent. Hence, if we wish the *Principia Mathematica* to be endowed with the property of always knowing how to give a proof of any theorem of arithmetic (i.e. to be **complete**), we will discover that there will be at least one case in which a theorem and its negation will arise, which means that the *Principia Mathematica* are not coherent. If, instead, we wish that the *Principia Mathematica* be coherent (that is, that it does not lead to contradictions), then we must accept, as a toll, that there will be at least a theorem on which it will not be possible to decide; i.e. *Principia Mathematica* are not complete. And, if someone insists on taking that theorem and raising it to the rank of **Axiom** (hence, of an **axiomatic truth** from which to start, together with the ones already given and accepted), then another, new theorem would arise for which one would repeat the discovery that it is impossible to decide. Gödel did all this by using numbers instead of formulae, in summary. Starting from the problems which afflicted the mathematical world (an example being Goldbach), Russell and Whitehead produced the illusion that the elimination of words and the use of only formulae was the solution of all troubles. Gödel, when introducing the numbers to replace the formulae, in 1931 discovered that the principle of the **excluded third**, after thousands of years, had to fall.

Gödel, one of the greatest mathematicians of the twentieth century, discovered that at the heart of the most rigorous mathematical logic – arithmetic – the "principle of the excluded third" is not valid. As said already (in Appendix 13) this principle had been held to be valid for thousands of years. His discovery is still fascinating today. It is not true that a proposition can be only true or false; it may be impossible to decide whether a given proposition is true or false.

Eugene Paul Wigner (1902–1995) was one of Gödel's few friends (see "*Wigner, von Neumann and Gödel*", Appendix 14). Wigner is the father of the famous theorem which establishes the existence of the Time-reversal operator. All laws of physics should be, as we now say, "T invariant", i.e. fundamental physics laws should remain the same if we invert the time axis: instead of going from past to future, we should also be able to go from future to past. Not us, but "elementary" particles. They have no memory, no clocks. Wigner is one of the greatest scientists of all times. He was able to demonstrate that the T-operator does not produce any contradiction in the Logic of Nature. But in 1964, a special effect was discovered which violates "T invariance" in the decay of a particle called K_L-meson (long-lived K-meson). This discovery attracted the attention of the physics community on an experiment done in the middle of last century and already mentioned on page 10 (Chapter I). My group gave the experimental proof that, in Quantum ElectroDynamics, T invariance is indeed valid. This is how I first met Wigner who invited me to Princeton to discuss the problem of T invariance.

Since Wigner was one of Gödel's few friends I asked him to let me know his friend who was working on the three pillars of our immanentistic (see "*Immanent and Transcendent*", Appendix 15) sphere: Language, Logic and Science. Gödel was interested in studying the creativity in mathematical logic (see "*Creativity in mathematical logic: Infinity*", Appendix 16), which has only one condition to satisfy: non contradiction. A fascinating mathematical structure that leads to no contradiction is the **Infinity**, which Gödel and Paul Cohen were able to demonstrate that it is not contradictory in any way. For creativity in Science the non contradiction is a necessary but not sufficient condition. It is needed the experimental reproducibility introduced by Galileo Galilei for first level Science, as emphasized by Blackett whose discovery of the 1st virtual effect (vacuum polarization) was known to Gödel, since he was a physicist of the famous "Vienna Circle" (Appendix 13: *"The Vienna and Gödel's discovery"*), as Blackett pointed out to me when Russell had expressed his judgement on us physicists.

Back to the Era of Peace which now days has become reality. Blackett and Russell wanted this but strongly disagreed over the ways and means to achieve it. We are now at the **crossroads** – will the political powers of free and democratic countries follow the road predicted by Russell's pessimism and feared by all the free worlds' scientists?

Here is the Russell view. "**The further away** a scientific discovery is from what the culture believes to have understood, **the more unpredictable the technological applications**, for good or evil, will be. You are talking now about fundamental forces and elementary particles without having understood what is at play. After the Nuclear Universe, you are working to discover another type, more sophisticated and microscopic, of Universe.

The governments, **as long as the secret laboratories continue to exist**, will keep financing them out of the reciprocal fear of the new technologies which could turn out to be applicable to arms that would be ever more precise and ever more potent."

Is this going to be the choice of the political powers? We must hope that the governments will go the way Blackett showed them during the difficult times following the end of the 2nd World War, by putting in place the **first European Scientific Institution, CERN,** which he wanted to be beyond all ideological, political and racial barriers, without secrets or frontiers, and dedicated entirely to fundamental research without any kind of war-like technological applications. Chien Shiung Wu [20]: "*Without the engagements of P.M.S. Blackett and I.I. Rabi, CERN would not have existed.*"

How could I interact so well with all members of the Blackett group despite my broken English? The answer is: Lady Costanza Blackett. The most difficult interactions were with Russell. In all discussions with Russell, the person who helped me with my broken English was Lady Costanza Blackett, who was fluent in Italian. She had an English mother and an Italian father. She spent all her youth in Florence. As a child Costanza nicknamed herself "Pat": Blackett and Costanza married in March 1924 and, to intimate friends, they were the two "Pats". She was very happy that her husband's name Patrick needed Pat, to start with. The two Pats were said to be the handsomest, gayest, happiest pair in Cambridge. The happy atmosphere in their house was the result of their profound love. When I became a member of the Blackett group, I could realize how important it was for me to have this charming Lady helping me not only in all language difficulties but also in encouraging my scientific activity, telling me how much her husband appreciated my knowledge on the new frontiers of physics and telling me that I was the new pupil of her husband. After more than half a century my gratitude for Lady Costanza Blackett is at the most profound level in my heart.

V – THE "BLACKETT EFFECT" IN THE 2nd WORLD WAR

Professor Blackett was Chief Advisor for Operational Research in the Admiralty during the 2nd World War.

The Blackett effect (**Figure 27**) refers to the problem of going from A to B. The belief of the Navy at the time was that the best solution is to send one ship at a time. It is hard to find a single ship in a sea to attack, as anyone who has played *Battleships* knows. If you send many ships it is easy to find many ships going together and attack all of them. But contrary to the Navy doctrine at the time, the best solution is to send as many ships as possible distributed in concentric circles, with the correct position of the escort vessels and air escorts in relation to the number of ships in the concentric circles.

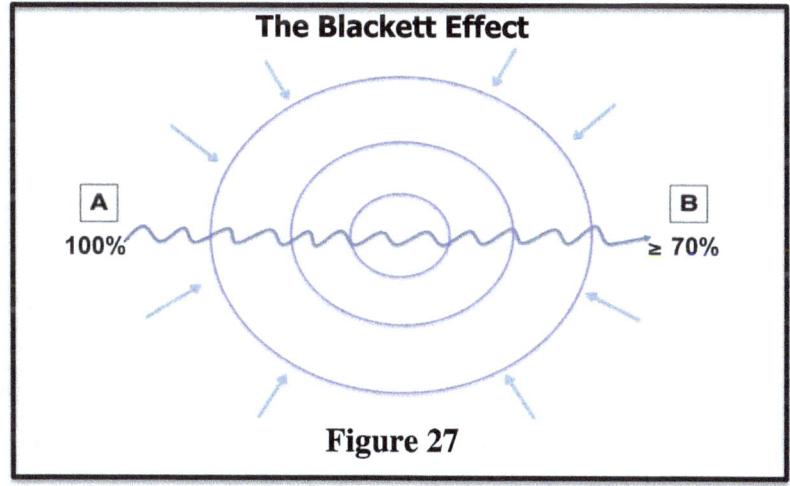

Figure 27

The maximum loss, as Blackett demonstrated, using the Probability Theory on the data of convoys of Allied merchant ships (see later), cannot exceed 30%. This was proven when the British Navy went on to liberate Malta with just a few per cent losses. The surrender of the Island (Malta) was considered inevitable as reported by British newspapers and Italian newspapers. I was a young fellow reading *"Corriere dei Piccoli"*. In this magazine for children it was written that if the British Navy attempted to liberate Malta they would be severely beaten.

No one knew the **"Blackett effect"**. And nobody knew that a great British mathematician, Alan Turing, had been able to decipher the Nazi's secret code. Using his discovery, the British let the Nazis believe that they could decipher the British secret messages. These messages were saying that the Allies would attack Sicily from the beach of Marsala (**Figure 28**) where they were planning to land the allied troops. Marsala beach was the same place where Garibaldi, in the previous century, started to liberate the Kingdom of the Two Sicilies. Needless to say that the correct plan was to start from Catania (**Figure 28**), which is on the opposite side of the island, something like 300 km away from Marsala.

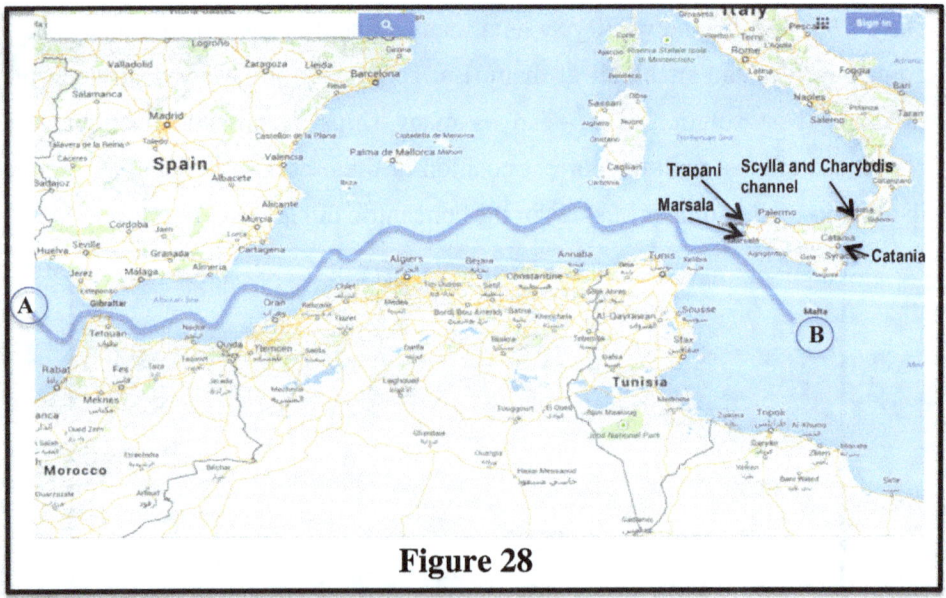

Figure 28

Many months before this great event, my father, who was an antifascist, after listening to London-Radio News decided to bring the family out of

Trapani (**Figure 28**), to escape the heavy bombing of the city where we were living. The countryside chosen was a place full of olive trees, in a region between Marsala and Trapani. One day hundreds of Nazis heavy tanks came. No one knew why. The only clear point was that the olive trees were excellent in hiding the enormous number of heavy tanks, avoiding detection from the air by the Allies. The full answer came a few days later when the British Navy started to bombard Marsala as if the Allies wanted to start from there to liberate Sicily. I could see the terrible scenario of the bombing of Marsala. My father was very sorry to have chosen this countryside. After three days of bombardment, all Nazi heavy tanks started to leave. They were not Ferrari cars and were only able to reach the opposite side of Sicily after three days, by which time the Allies were able to be many kilometres inside. The Nazi heavy tanks were given the order to abandon Sicily. The danger for the Nazi army was to be isolated from the continent due to the Scylla and Charybdis channel (**Figure 28**).

The British Navy was able to win the Mediterranean battle, during the 2nd World War, thanks to the "**Blackett effect**".

This great victory was so unexpected and so great that the Allies decided to change their strategy and immediately started to liberate Sicily where the war ended in 1943, averting two years (July 1943 – April 1945) **of terrible tragedies in the place where I was born.**

To recall how the "**Blackett effect**" could be achieved is of great value for the future of our civilization: it is an excellent example of how Science can help to solve problems often considered not to be there, despite the terrible consequences being in front of us.

A few examples. From the simplest (the colour of the aircraft) to the less simple ones, like the application of Probability Theory for the defence of shipping lanes.

The statistical analysis of the data on the speed and size of convoys of Allied merchant ships in relation to numbers of ships lost is how the "**Blackett effect**" has been obtained.

Blackett joined the instrument section of the Royal Aircraft Establishment few months after the outbreak of the war. This was his first step in bringing Science in military actions and strategy. The second step was the Coastal Command.

In Coastal Command operations the aircrafts were easily spotted by Nazi-submarines because they were dark objects against a bright sky.

Blackett convinced the Coastal Command to paint antisubmarine aircraft **white** instead of **black**. Bomber aircraft were painted **black** in order to reflect as little light as possible from **searchlights**. Blackett proved the point by comparing photographs of **crows** and **seagulls** against different skies. The mathematical result was an increase of 30 per cent in sinkings of the Nazi-U-boats. This prediction was confirmed by the new data of sinkings of Nazi-submarines having only changed the colour of the aircraft from black to white.

The next result is the following. When a U-boat was spotted, the existing methods of attack gave a few per cent kill-rate. This very low effect seemed to be due to the power of the explosive. The power of the explosive could not be increased since it was the highest available. The problem was not in the low power of the explosive but in the low bombing accuracy. Blackett succeeded in improving the depth adjustment.

From captured German U-boat crews Blackett learned that they thought the UK had discovered a new and very powerful explosive. During a conversation with his young fellows at Imperial College he told us "Actually we had only turned a depth-getting adjuster from 100 foot to the 25 foot mark." The U-boat sinking from 3 per cent went up to 40 per cent. There can be few cases where a factor ten gain can be obtained by such a small and simple change.

A much less simple recommendation to the Coastal Command was to use Probability Theory to defend shipping lanes. The Coastal Command had positioned ten fighters, for the defence of the "West of Ireland shipping lanes". The problem was that the Coastal Command was convinced that the use of the ten fighters could "sweep clean" the German area of operation. Blackett considered fallacious the argument that when the ten aircrafts were flown, the whole area was swept clean so that the chance of sighting the enemy was 100%, but when any fewer number of fighters was flown, the chance of sighting the enemy was zero.

There are many more than these two cases, and they can be modeled using the mathematics of Poisson distribution. **The probability** that German aircraft would be in the area on the day when all ten fighters were up, **and all the other probabilities** that the German aircrafts would be in the area when fewer fighters fly every day, **are** all given by the **Poisson distributions**. The correct answer is given by **summing small probabilities** and the results show that the sum won against the arguments for gambling on an occasional certainty. Whatever fighters were available had to be flown. No matter what the day, try to fly every day whatever fighters are available.

And now comes the biggest step. Sir Charles S. Wright, Director of Scientific Research at the Admiralty, asked Blackett in **December 1941**, to explain the methods that he was using in OR (Operational Research) at the Coastal Command and to write a memorandum on the same topic.

Blackett's presentation and his memorandum impressed the Admiralty so much that, just one month later, in January 1942, he was asked to transfer to the Admiralty. This is how he became the Director of Naval Operational and Research. His memorandum "Scientists at the Operational Level" circulated widely in UK and in the United States.

The **first priority** in his new responsibility was to analyse with scientific methods the speed and size of convoys of Allied merchant ships, escort vessels, and air escorts in relation to numbers of ships lost.

A surprising result of statistical analysis of data from the previous two years was that forty ships were safer than smaller groups of ships, contrary to Navy doctrine. This is how the "**Blackett effect**" was discovered and the two years of peace instead of the two terrible years of war became real life in Sicily.

The "**Blackett effect**" and the contribution of Alan Turing in the 2nd World War have been forgotten while considered – as previously emphasized – by the physicists of the Blackett Lab very instructive: an excellent example of how Science can help to solve problems often considered not to be there despite the terrible consequences being in front of all of us.

VI – BLACKETT, THE CAMBRIDGE CIRCLE AND THE WHOLE OF OUR KNOWLEDGE INCLUDING VIRTUAL HISTORY AND THE THREE BIG BANGS

> *On the occasion*
> *of my 1st anniversary in the Blackett group,*
> *Professor Blackett*
> *gave me a photo taken*
> *the same year in which I was born.*
> *In this photo,*
> *there are the authors*
> *of nearly*
> *all of our knowledge in 1929.*

A few words (a special volume would be needed) for each of these great scientists.

– 1929 –

Photo 9
(From left) Patrick Maynard Stuart Blackett, Pyotr L. Kapitza, Paul Langevin, Ernest Rutherford and Charles Thomson Rees Wilson outside Cavendish Laboratory (1929). **This photo – dated 1929 – was a gift in 1955 from Professor Blackett to the youngest member of his group** (A.Z.). On this occasion, Professor Blackett said that in the same year (1929), the best synthesis on the future of physics had been given in Florence by **Orso Mario Corbino**, the founder of what would have been the famous "Panisperna Group" where Enrico Fermi invented the "slow neutron technology" which allowed all elements of the Mendeleev Table to become "radioactive". This was how Fermi discovered the "weak forces". In the words of Professor Corbino[(*)]: *"The only possibility of great discoveries in physics dwells in the eventuality of succeeding in modifying the inner nucleus of the atom. This will be the task truly worthy of the future in physics."*

(*) *"La sola possibilità di grandi scoperte in Fisica risiede nella eventualità che si riesca a modificare il nucleo interno dell'atomo. E questo sarà il compito veramente degno della Fisica futura."*

Blackett – the discovery of (e⁺e⁻) production in cosmic rays and of the first examples of strange particles: a mesonic state and a baryonic state. This is why the "mesonic state" (θ^0) and the "baryonic state" (Λ^0) were given opposite values of the new quantum number called "strangeness": s = **+1** and s = **−1**, respectively. The discovery of "mesonic states" with positive and negative strangeness came in 1957.

Kapitza – the discovery of superfluidity (and no secret Labs, Erice Statement); he was the hero of science in the USSR for his courage to decline the directorship of the nuclear-fusion bomb (now called the H–bomb) thousands of times more powerful than the nuclear-fission bombs that destroyed Hiroshima and Nagasaki.

Langevin – the discovery of paramagnetism and the invention of the "twin paradox". He gave to Becquerel a suggestion which led to the discovery of radioactivity.

Rutherford – the discovery of the atomic nucleus, i.e. of the fact that atoms are composite objects and that more than 99% of their mass is concentrated in a volume which is a millionth of a billion times smaller than the atomic volume.

Wilson – the invention of the "**cloud-chamber**": the instrument which allowed, for more than five decades, the enormous number of discoveries using "cosmic rays" as primary high energy particles.

The 1929 photo (**Photo 9**) was taken by the youngest member of the "*Cambridge Circle*", Paul Dirac (this is why he is not in the photo). In the "*Cambridge Circle*" there were three components: the "*Blackett Circle*", the "*Dirac Circle*" and the "*Kapitza Circle*", each one following the main physics and scientific interest of his leader.

Here is what Professor Blackett was teaching to us young fellows.

> *The fantasy of nature*
> *is much more powerful*
> *than the fantasy of all of us.*

In fact: "*Nature has been very kind with us: all new particles have a lifetime of $\simeq 10^{-10}$ sec (perfect for centimetres flights). The production time for the same particles is $\simeq 10^{-23}$ sec.*"

In 1897, when the first "elementary" particle was discovered by J.J. Thomson, no physicist could have imagined the enormous number of "elementary" particles that would be discovered in the next century. No physicist could have imagined that all these particles would not have "scalar" properties in the Lorentz Space-Time and in the "intrinsic" spaces, which were later found to exist.

The apex of all these findings came in 1964 when the introduction of the imaginary mass in the Lagrangian (energy density) was proposed. The existence of an imaginary mass came after a sequence of "**U**nexpected **E**vents with **E**normous **C**onsequences" ("UEEC") (see later and **Figure 29**) related to the totally unexpected existence of so many "elementary" particles, all coming after the discovery of the "smallest piece of electricity" named the "electron" in 1897 (see "*Why electrons are so important*", Appendix 17).

The name given to this new particle, needed as a consequence of the introduction of the imaginary mass in the Lagrangian, was S^0. At that time no one knew that the S^0 would be called the "God particle".

I was many decades younger than now when the discovery of the S^0 particle (S ≡ Scalar in all spaces: Lorentz and intrinsic; zero indicates the value of the electric charge) was announced by an American group in 1965 [99].

It was the first time that the existence of a particle with all zero quantum numbers (zero electric charge, zero Lorentz spin, zero spin in all internal symmetries) was experimentally proven. During 50 years, from 1897, when the first elementary particle (electron) was discovered, to 1947, only five other particles were discovered to exist, for a total number six (e, γ, p, n, π, ν). **From 1947 to 1967, during two decades, a totally unexpected large number of particles were discovered: all with at least one non-zero intrinsic quantum number.**

The S^0 was the first candidate for **what later became known as the "God particle"**. It was the era of bubble-chamber dominance and the NBC (non-bubble-chamber) physicists were invited to corroborate this first great discovery using NBC technology. **The S^0 discovery attracted everybody's attention.**

Remembering Professor Blackett's instructions, **an exact repetition of the experiment was needed**: the same energy, the same solid angle and all other essential details. We had at CERN a powerful NBC set-up and **the results found did not show any evidence for the existence of the S^0–meson** [100]: **in disagreement with those where the S^0 was discovered** [99]. The group leader came to visit my experimental set-up and, after three days of analysis, declared our experimental results to be correct.

In the next 45 years no-one was able to find the scalar meson with all zero properties. Leon Lederman devoted all the efforts of the Fermi Lab and an enormous amount of time to prove its existence. The enormous difficulty for its discovery **is the only justification** for the name "**God particle**" given to the **boson of Brout-Englert-Guralnik-Hagen-Higgs and Kibble** (see references quoted below).

The "God particle" was discovered at CERN in July 2012.

My warmest wishes to my theoretical friends (Englert, Brout, Higgs, Guralnik, Hagen, Kibble) for their theoretical work, to all the CERN experimental physicists who have contributed to the discovery of the "God particle" in particular to Fabiola Gianotti spokesperson of ATLAS and to Joe Incandela spokesperson of CMS; to Rolf-Dieter Heuer, CERN Director General, and to the CERN Research Director, Sergio Bertolucci.

This discovery has allowed us to have the theoretical description of the Logic of Nature with non-zero masses. As mentioned before, this problem goes back to 1964 when Brout, Englert, Guralnik, Hagen, Higgs and Kibble gave to the physics community their proposal to solve the problem by introducing in the virtual physics phenomena the imaginary mass for the theoretical description of the Logic of Nature.

Article	Reception date	Publication date
F. Englert and R. Brout *Phys. Rev. Letters* **13** *(1964) 321*	26/06/1964	31/08/1964
P.W. Higgs *Phys. Letters* **12** *(1964) 132*	27/07/1964	15/09/1964
P.W. Higgs *Phys. Rev. Letters* **13** *(1964) 508*	31/08/1964	19/10/1964
G.S. Guralnik, C.R. Hagen and T.W.B. Kibble *Phys. Rev. Letters* **13** *(1964) 585*	12/10/1964	16/11/1964

VIRTUAL PHYSICS AND VIRTUAL HISTORY

The historians have invented "Virtual History" following the physics discovery of "Virtual Physics".

As said before, Professor Blackett is the physicist who discovered, in 1932, the simultaneous production of (e^+e^-) pairs in cosmic rays, thus giving the necessary experimental evidence for the existence of the first example of "Virtual Physics", the "vacuum polarization effects", theoretically predicted by Dirac in 1929.

"Virtual **Physics**" and "Virtual **History**" operate in the two asymptotic limits of complexity (minimum and maximum levels) but share a common property: **Evolution**.

In fact History is "**E**volution of the **W**orld in its **R**eal **L**ife" ≡ (**EWRL**), and Science is "**E**volution of our **B**asic **U**nderstanding of the laws governing the world in its **S**tructure" ≡ (**EBUS**).

The two evolutions (EWRL) and (EBUS) are dominated by **U**nexpected **E**vents with **E**normous **C**onsequences (**UEEC** events) as illustrated in **Figure 29**.

The world would not be as it is if any of the UEEC events reported in the History column of **Figure 29** did not take place. The virtual world, which would come from "Virtual History", would not be like our world, but "Virtual Physics" is exactly our physics.

"Virtual History" could not have been invented if "Virtual Physics" had not been discovered, thanks to Professor Blackett, in 1932.

In **Figure 29**, Science and History are compared. The right side of **Figure 29** will be reproduced in **Figure 36** when we will discuss the UEEC in Science.

The Logic of Nature allows the existence of the lowest limit of Complexity, **Science**, and of the highest limit of Complexity, **History**. How to go from the lowest to the highest limit of Complexity is illustrated in **Figure 30**.

UEEC IN HISTORY AND IN SCIENCE

	In History = EWRL		In Science = EBUS
I	What if Julius Caesar had been assassinated many years before?	*I*	What if Galileo Galilei had not discovered that $F = mg$?
II	What if Charles VII had not been able to win the 100 years war?	*II*	What if Newton had not discovered that $$F = G\frac{m_1 \cdot m_2}{R_{12}^2} \; ?$$
III	What if America had been discovered a few centuries later?	*III*	What if Maxwell had not discovered the unification of electricity, magnetism and optical phenomena, which allowed him to conclude that light is a vibration of the EM field?
IV	What if Napoleon had not been born?	*IV*	What if Becquerel had not discovered radioactivity?
V	What if Louis XVI had been able to win against the 'Storming of the Bastille'?	*V*	What if Planck had not discovered that $h \neq 0$?
VI	What if the 1908 Tunguska Comet had fallen somewhere in Europe instead of Tunguska in Siberia?	*VI*	What if Lorentz had not discovered that space and time cannot both be real?
VII	What if the killer of the Austrian Archduke Franz Ferdinand had been arrested the day before the Sarajevo event?	*VII*	What if Einstein had not discovered the existence of time-like and space-like real worlds? Only in the time-like world, simultaneity does not change, with changing observer.
VIII	What if Lenin had been killed during his travelling through Germany?	*VIII*	What if Rutherford had not discovered the nucleus?
IX	What if Hitler had not been appointed Chancellor by the President of the Republic of Weimar Paul von Hindenburg?	*IX*	What if Hess had not discovered cosmic rays?
X	What if Pyotr Kapitza accepted to be the leader of the USSR H-bomb Project as wanted by Stalin?	*X*	What if Dirac had not discovered his equation, which opens new horizons, including the existence of the antiworld?
XI	What if Nazi Germany had defeated the Soviet Union?	*XI*	What if Fermi had not discovered weak forces?
XII	What if Karol Wojtyla had not been elected Pope, thus becoming John Paul II?	*XII*	What if Fermi and Dirac had not discovered the Fermi–Dirac statistics?
XIII	What if Gorbachev had not been defeated by Yeltsin?	*XIII*	What if Yukawa had not proposed the existence of a "meson" in order to have the nuclear glue?
XIV	What if the USSR had not collapsed?	*XIV*	What if the 'strange particles' had not been discovered in the Blackett Lab?

Figure 29

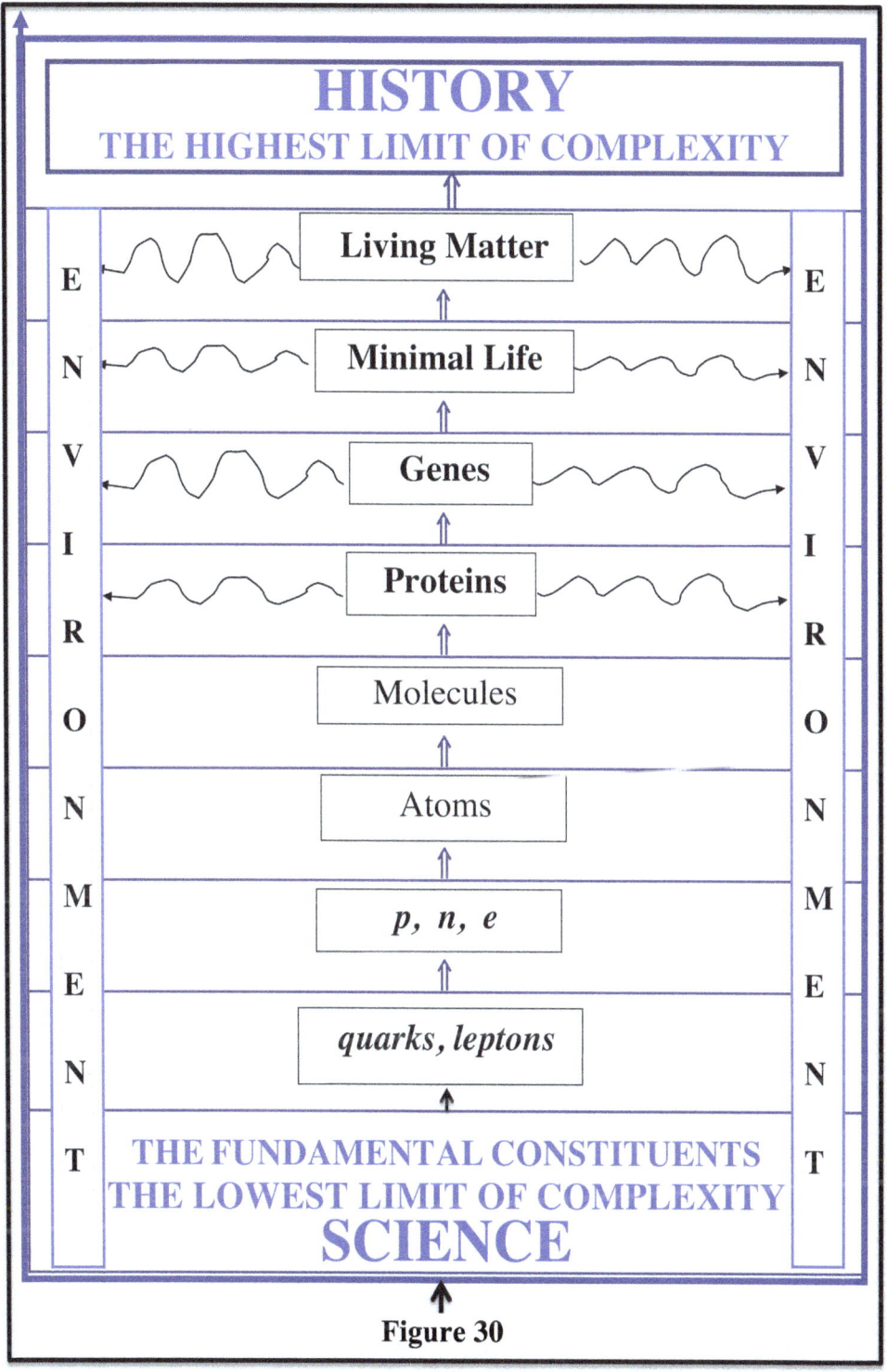

Figure 30

Science and History bring us to consider the "whole of our knowledge", reported in **Figure 31**. A few words on **Figure 31**.

The **First Big Bang** (*BB1*) is needed in order to describe how we go from the Vacuum to the Universe, made of inert matter.

The **Second Big Bang** (*BB2*) is needed in order to describe how we go from inert matter to matter endowed with life. This is being studied in many laboratories. Hundreds of scientists are fully engaged in studying what is called the "problem of minimal life", i.e. how many pieces of inert matter are needed in order to produce the most elementary piece of "living matter" (see "*The problem of going from inert matter to living matter*", Appendix 18).

The **Third Big Bang** (*BB3*) is needed in order to describe how we go from matter endowed with life (and no Reason) to the most elementary form of living matter endowed with the privilege of also having Reason.

Please note that the term Reason is not referring to the most simple form of Reason needed by living matter in order to guarantee life. The term Reason is the form of Reason which produces Language, Logic and Science.

There are hundreds of thousands of forms of living matter (vegetal and animal, small and big) but only one form of living matter is able to invent "**Language**" (which gives rise to Permanent Collective Memory, better known as Written Language), "**Rigorous Theoretical Logic**" (known as Mathematics) and "**Rigorous Experimental Logic**" (known as first level Galilean Science).

These three Big Bangs are indicated in **Figure 31**, where the knowledge of the Universe plays a central role.

However, if the Third Big Bang had not occurred, none of us would have been able to discuss our problems and I could not have written this book. The content of this chapter and of **Figure 31** includes many points I discussed with Professor Blackett when I was the youngest member of his group. Let me try to give a recollection of these interesting discussions.

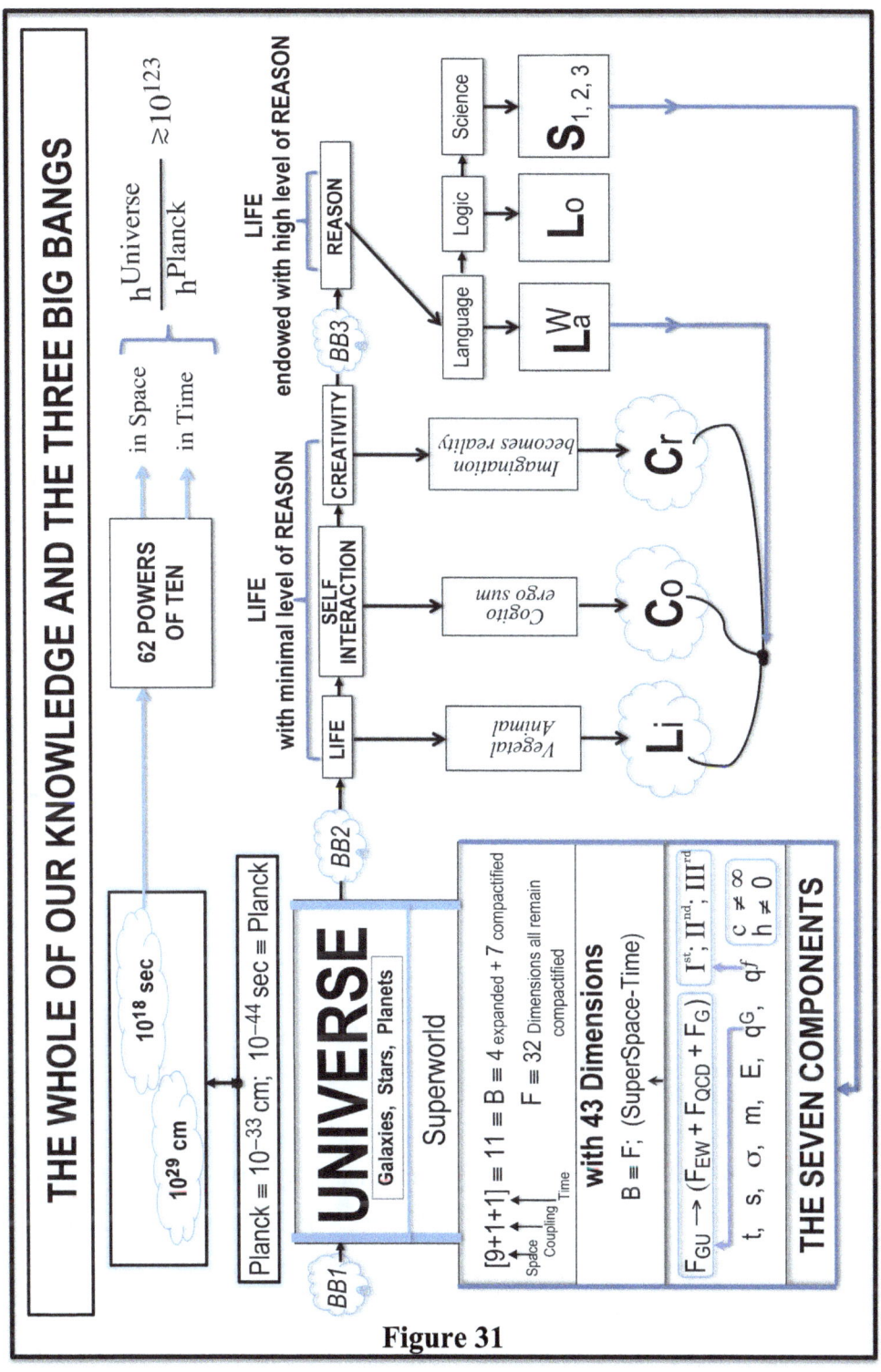

Figure 31

It is thanks to Science that we have discovered the seven fundamental components needed to construct the real world as we know it and participate in it. These seven components are **space** (s), **time** (t), **spin** (σ), **mass** (m), **energy** (E), **gauge charge** (q^G) which generates the Fundamental Forces of Nature and **flavour charge** (q^f) which produces the stability of all fundamental particles, called quarks and leptons.

The seven components are the result of the three Big Bangs. The description of **Figure 31** would take too much space. We will limit ourselves to the shortest possible remarks.

The seven quantities must give rise to quarks and leptons which exist into three "Families". This is the meaning of I, II, II in **Figure 31**. The seven quantities must also generate the four fundamental forces (see *"Fundamental Forces"*, Appendix 2). Notice in **Figure 31** the Electromagnetic and the Weak Forces are mixed into the so called Electroweak Force (F_{EW}). The reason being that these two forces start at the Fermi Energy from a mixed source. The other two forces are, the Subnuclear Strong (F_{QCD}) and the Gravitational Force (F_G). All these forces should be unified into the Grand Unified Force (F_{GU}).

As discussed in Appendix 4, *"Fermions and Bosons"*, all fundamental particles are not balls but spinning balls with semi-odd-integer value for their spin. All glues produced by the fundamental forces are also spinning balls, but with integral value for their spin. The spinning balls with semi-odd-integer value for their spin

$$\left(1/2 \; ; \; 3/2 \; ; \; 5/2 \; ; \; 7/2 \; ; \; \ldots \right)$$

are called "Fermions", **F**. The spinning balls with integer value for their spin

$$(\; 0, \; 1, \; 2, \; 3, \; 4, \; \ldots \;)$$

are called "Bosons", **B**. Fermions obey a statistical law totally different from the statistical law which have to obey the Bosons. The fermionic statistical law establishes that two identical Fermions cannot be in the same place. The bosonic statistical law establishes that in the same place can be any number of Bosons.

The "same place" is a simple way to specify not only a "place" but a series of properties needed to identify an elementary particle.

The Superworld would be the result coming from a new Symmetry between Fermions and Bosons, (B ≡ F). This "**Supersymmetry**" should be valid not only for "particles" but also for Space and Time.

This is the origin of Superspace-Time with 43 dimensions, which are 11 bosonic and 32 fermionic. The 11 bosonic dimensions are 9 for Space, one for the fundamental coupling and one for Time. Out of the nine bosonic dimensions of Space, six remain compactified and three are expanded. The other bosonic dimension which is expanded is the Time dimension. The total number of bosonic expanded dimensions is therefore four.

In our world we have in fact four bosonic expanded dimensions: three for Space and one for Time. This is why we have so much **Space** and so much **Time** available. A volume of our Space needs three dimensions: height, width and length. The fact that all these three quantities can be measured with the same instrument, called meter, is the proof that our Space has three dimensions. The dimensions needed for Time is only one and is measured with a clock (Einstein dixit).

Along the Time dimension there are phenomena for which the two arrows of Time (from past to future and from future to past) are allowed, thus obeying to the Wigner theorem (discussed in Chapter IV). Others phenomena are not obeying the Wigner theorem. In synthesis: only some "elementary processes" can go in the two opposite directions along the Time axis without changing any of their properties.

The other seven bosonic dimensions remain compactified. As said already, one of these corresponds to the fundamental coupling of the Grand Unified Force (F_{GU}) from which all other forces (F_{EW}; F_{QCD}; F_G) are generated.

In addition to the 11 bosonic dimensions there are 32 fermionic dimensions. These remain all compactified. As mentioned before, the bosonic and fermionic dimensions are the basis of the Superspace with a total of 43 dimensions.

An important property distinguishes the fermionic dimensions from the bosonic dimensions.

In the Superspace with bosonic dimensions it is possible to go in one direction or in the opposite direction, as we do in everyday's life along the three expanded dimensions of Space. In the Superspace with fermionic dimensions this is impossible.

The Superspace with 43 dimensions, and all their properties, which we have briefly described, are needed for everything that refers to inert matter. Life extends beyond the confines of inert matter. When we introduce life into a description of the world, the first problem to solve, as already mentioned, is the transition called Big Bang Two ($BB2$) from inert matter to living matter.

There are very many forms of matter with life (vegetal and animal) and with very low levels of Reason. The number of all these forms of life is in the range of a million. Despite this enormous number, only one form of living matter has the privilege of being endowed with the extraordinary property of a high level of Reason. As shown in **Figure 31** the Third Big Bang ($BB3$) is needed in order to have the transition to this form of matter. It is thanks to the existence of matter with high level Reason that we have been able to discover **Language**, **Logic** and **Science**. Language has produced "Permanent Collective Memory" (PCM) better known as "writing". With logic we mean Rigorous Theoretical Logic, better known as Mathematics. There is another form of Rigorous Logic, which needs experimental reproducibility, and it is known as Science. Two more details. In **Figure 31** the letter **c** indicates the speed of light which is a fundamental constant of Nature and it is not infinite (∞) as believed to be for millennia. The first fellow who attempted to measure the speed of light is Galilei in the sixteenth century. He would have succeeded to prove that the speed of light was not ∞ if it would have been thirty times the speed of sound. The speed of light is not infinite but a million times greater than that of sound. This is why Galilei was unable to measure it. Galilei had at his disposition the distance between two hills in the Tuscan countryside, in other words just a few kilometres. We had to wait Ole C. Rømer (1644–1710), who used Jupiter's *Io* as the cosmic lantern, and the orbital velocity of the Earth. This gave him the "million times" factor that Galilei did not have. Not only was *Io* discovered by Galilei, but the regularity in the light signals emitted by the moons of Jupiter had led Galilei to propose using it as a "cosmic clock". And it was by using Galilei's "cosmic clock", and the speed by which the Earth moved, away from and closer to Jupiter, due to its orbital motion, that Rømer succeeded in demonstrating that the speed of light is not infinite measuring for the first time after Galileo Galilei, its finite value, as Galilei was expecting.

The other detail in the same box of **Figure 31** refers to the quantity h, which is the other fundamental quantity of Nature: action. For millennia this quantity was believed to be as small as wanted, including zero. Planck proved, in 1900, that the smallest quantity of action cannot be zero, $h \neq 0$, but greater or equal to the quantity now called Planck's action. If I give to a friend of mine a bit of energy for a small interval of time, this is a small action which corresponds to a very high number – billions of billions of billions – of the smallest amount of elementary action, i.e. the Planck's action.

Let us calculate the energy needed to bring one kilogramme of mass up by one meter. If this energy is multiplied by the Time of one second the result is a quantity of action equal to ten million of billion billion billion times the Planck's action.

A few other details for **Figure 31**. **Li** corresponds to all forms of matter endowed with Life, vegetal and animal. **Co** indicates all interactions that matter can have with life and with a minimum level of Reason. **Cr** stands for Creativity which is when even the lowest level of imagination becomes reality.

L_a^W is Written Language; **Lo** is the Theoretical Logic (Mathematics) and Science – as said several times – is Rigorous Experimental Logic. Science has three levels, this is the meaning of $S_{1,2,3}$. The 1st level corresponds to reproducible experiments in a Laboratory: for example, the discovery of Antimatter [101]. The 2nd level corresponds to observations with no possibility of intervention: example, the study of evolution of Stars. The 3rd level is when something happens only once: example, the first Big Bang. The 3rd level seems to be in contradiction with reproducibility and could seem in contradiction with the meaning of Science. All three Big Bangs are Science because their description can never be in contradiction with what has been discovered at the 1st level of Science.

In **Figure 31** the Planck length (10^{-33} cm) and the Planck time (10^{-44} sec) are given. These two quantities have been discovered by Planck in 1900 when he took as fundamental units the values of the three fundamental constants of Nature:

        1) the speed of light;
        2) the value of the Planck action;
        3) the Newton constant.

When the radius of the Universe (10^{29} cm) and the age of the Universe (10^{18} sec) are divided by the Planck length and the Planck time the result is 10^{62}:

$$\frac{10^{29}\,\text{cm}}{10^{-33}\,\text{cm}} = \frac{10^{18}\,\text{cm}}{10^{-44}\,\text{sec}} = 10^{62}.$$

These are two very meaningful big ratios linked to our world.

A statement concerning the biggest number. The biggest number comes out when the action of the Universe is divided by the Planck action. This ratio is the number "**one**" followed by hundred twenty-three zeros: 10^{123}, as reported in **Figure 31**.

Figure 31 recalls to me how much gratitude I must have towards Professor Blackett, whose interest for the whole of our knowledge has given me the intellectual stimulus for such a Figure.

VII – NEW INSTITUTIONS FOUNDED

Being a pupil of Blackett has been of vital importance not only for my career in physics but also for my other activities, including the establishment of the Ettore Majorana Foundation and Centre for Scientific Culture (**EMFCSC**) in Erice.

I recall that **Professor Blackett** was convinced that it is us, the physicists, who must be engaged directly to let the people outside our labs know what is Science so that they become aware of what the role of Science for the progress of our civilization is.

The scientific discovery at the 1st level of Galilean Science is the real motor for progress in technological inventions that allow the quality of life to be at the level it is today.

In the constitutive act of the EMFCSC (**Figure 32**) Professor Backett was one of the founders, together with

---

J.S. Bell, I.I. Rabi, V.F. Weisskopf and A.Z.

The Constitutive Act of the Erice Centre in the Blackett Institute

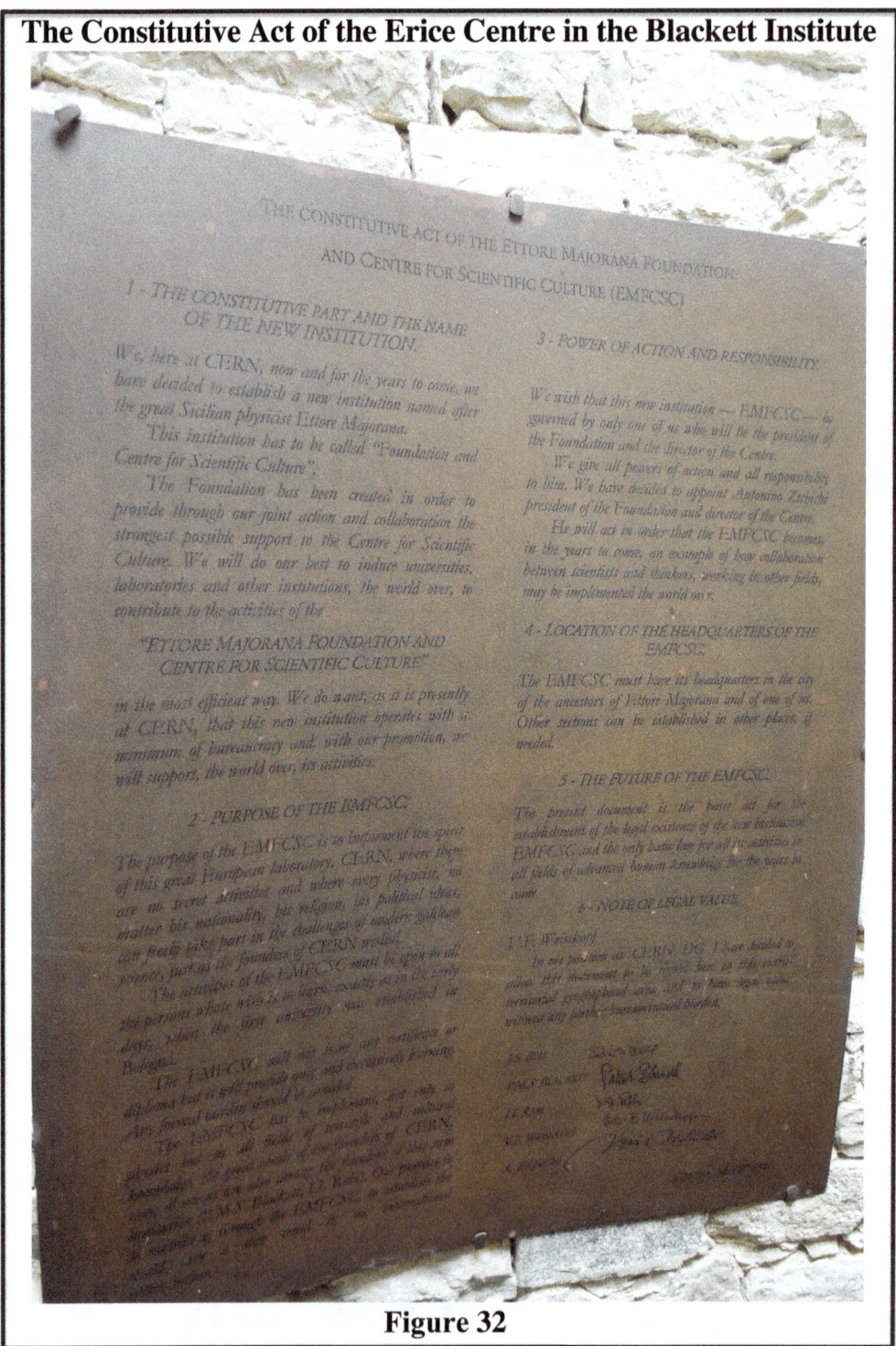

Figure 32

In 1991 the Director of the Blackett Laboratory at Imperial College in London, Professor Thomas W.B. Kibble, came for the Ceremony of the Inauguration of the Blackett Institute in Erice (**Figure 33**).

Figure 33: P.M.S. Blackett's Statement and the Director of the Blackett Laboratory at Imperial College in London, Thomas W.B. Kibble.

At the entrance of this Institute there is the famous Blackett statement:

> *"We experimentalists are not like theorists: the originality of an idea is not for being printed in a paper, but for being shown in the implementation of an original experiment."*
> Lord Patrick Maynard Stuart Blackett, 1962

This statement by Blackett was not directed against theoretical physics. The best proof is in the famous Feynman statement: *"It doesn't matter how beautiful your theory is, it doesn't matter how smart you are. If it doesn't agree with experiment, it's wrong."* Blackett wanted to be sure that we, young fellows, know what was boiling in the field of theoretical physics. But in the field of experimental physics, we had to know that it is the implementation of an original experiment the decisive step for the progress of our science (see *"The Ten Challenges of our Physics"*, Appendix 19).

In Geneva in **1985**, Reagan and Gorbachev declared that secret labs are the number one enemy of peace and so we will open all secret labs. **Since this dream turns** out to be a very difficult one to come true, the only way out is to destroy the secret **at its origin**.

The motor of progress is scientific discovery, which generates technological inventions and **here lies the problem**. Inventions can be both for peaceful and for war technology. We must stop the use of technology for war and build the most powerful machine in the world, the ELN, **E**uroasiatic **L**ong **I**ntersecting **S**torage **A**ccelerator (see Chapter III-3 and **Figure 13** plus Chapter X and **Figure 49**). If the Berlin Wall had not fallen, ELN would have been built since **Reagan, Gorbachev** and **Deng Xiao Ping** had all agreed (see **Photo 10**).

Photo 10: *From left to right (first row):*
Professor Zhou Guang Zhao (Scientific Advisor to Premier Deng Xiao Ping), Professor Edward Teller (Scientific Advisor to President Reagan), Professor Antonino Zichichi (Chairman of the International Committee "Science for Peace") and Professor Eugenij Velikhov (Scientific Advisor to President Gorbachev), shaking hands after reaching the Agreement for International Scientific Collaboration East-West-North-South without Secrecy and without Frontiers (1986).

And now a few words on the other Institutions which are the "analytical continuation" of Professor Blackett, being my strong supporter: the WFS and the ICSC-Word Laboratory, with a summary of their results and achievements.

THE WORLD FEDERATION OF SCIENTISTS
Organisation & Status

The World Federation of Scientists (WFS) was founded in Erice, Sicily, in 1973, by a group of eminent scientists led by Isidor Isaac Rabi and A.Z.. Since then, many other scientists have affiliated themselves with the federation, among them T.D. Lee, Laura Fermi, Eugene Wigner, Paul Dirac and Pyotr Kapitza.

The WFS is a free association, which quickly grew to include more than 100,000 scientists drawn from 140 countries. All members share the same aims and ideals and contribute voluntarily to uphold the federation's principles.

The federation promotes international collaboration in science and technology between scientists and researchers from all parts of the world – North, South, East and West.

The federation and its members strive towards an ideal of free exchange of information, where scientific discoveries and advances are no longer restricted to a selected few. The aim is to share this knowledge among the people of all nations, so that everyone may benefit from the progress of Science.

The founding of the WFS was made possible by the existence, in Erice, of the Institution mentioned many times: the Ettore Majorana Foundation and Centre for Scientific Culture (EMFCSC), named after the physicist Ettore Majorana (see *"Ettore Majorana: Genius and Mystery, what Robert Oppenheimer told me"* Appendix 20).

This centre, which has been dubbed "The University of the Third Millennium", has attracted (as reported on page 6) over 100,000 scientists from all over the world since its founding in 1963. The EMFCSC was a precursor of the WFS and its function is to mitigate planetary emergencies.

The WFS rapidly identified 15 classes of planetary emergencies and began to organise the fight against these threats. The photo below shows one of the first meetings.

Photo 11: From left: Luigi Dadda, Pierre A. Piroué, Enrico Bignami, Yuval Ne'eman, Richard L. Garwin, John C. Eccles, Eugene P. Wigner, A.Z., Edward Teller, Paul Adrien Maurice Dirac, George Charpak (1981).

One of its main achievements was the drawing up of the Erice Statement, in 1982 (see page 137). The statement clearly sets out the ideals of the federation and puts forward a set of proposals for putting these ideals into practice.

Another milestone was a series of International Seminars on Nuclear War which have had a tremendous impact on reducing the danger of a planet-wide nuclear disaster, and ultimately contributed to the end of the Cold War.

In 1986, through the action of a group of eminent scientists (most of whom were members of the WFS) the International Centre for Scientific Culture (ICSC)–World Laboratory was founded in Geneva to help achieve the goals outlined in the Erice Statement. To achieve this, specific pilot projects have been implemented to overcome the planetary emergencies.

The ICSC–World Laboratory works on the principle that one of the better ways of helping developing countries is to support the participation of their scientific elite in projects aimed at the solution of their particular problems, working in collaboration with their peers in developed countries and contributing to the advancement of science and human knowledge as a whole.

Other achievements were the establishment of the Erice Prize, the Gian Carlo Wick Gold Medal Prize, the formulation of the Farnesina Statement, the Lausanne Declaration and the National Scholarship Programmes.

THE ICSC – WORLD LABORATORY
Geneva – Moscow – Beijing – New York

What is the World Laboratory

The International Centre for Scientific Culture (ICSC)–World Laboratory, is an international non-governmental organisation recognised by the United Nations and granted, by the federal government of Switzerland, the status of "Public Utility Organisation".

A total of 20 agreements were signed with governments from developing countries.

A total of 42 World Laboratory Research Centres were established in developing countries with the collaboration of national institutions.

A total of 65 scientific collaboration agreements were signed with governmental institutions and research institutes.

It has successfully implemented over 100 pilot projects, in all scientific fields, in developing countries and granted access to higher institutes of learning and research laboratories to more than 5,000 young scientists from developing countries. The World Laboratory is able to obtain the scientific cooperation of its members and project directors on a totally voluntary basis.

A Summary of Results and Achievements
(in collaboration with the World Federation of Scientists)

One hundred pilot projects and dozens of programmes were dedicated to the fight against planetary emergencies. Some results follow:

Basic Science and Technology –

Hundreds of bright young scientists from developing countries were given training at CERN, to help in the development of highly sophisticated detectors for the LAA, the ALICE and the R&D for Advanced Technologies experiments.

National Scholarship Programmes were established in 38 developing or newly emerging countries, with over 2,000 scholarships granted every year.

An International Scholarship Programme, in association to the Pilot Projects, granted over 1,000 international scholarships to bright young scientists from developing countries, for training in the best institutes and laboratories in industrialised countries.

Young scientists from developing countries were involved in the development of frontier research on spectroscopy equipment in our research centre in Uppsala.

A large number of young scientists from developing countries were involved in the cosmic rays research programmes in the Gran Sasso underground laboratory.

A centre for advanced science and technology was established in Beijing, in collaboration with the Chinese Academy of Sciences. It was the first established National Scholarship Programme and, over the last 18 years, it has consistently introduced whole frontiers of science areas in China, with 484 one-month-long domestic symposia and 18 international symposia training more than 30,000 scientists; 168 volume publications and nearly 4,150 papers published in international journals.

A research centre for high energy physics and cosmology was established in Islamabad.

A computer-assisted design centre was established in the University of Buenos Aires and training was provided to young scientists.

A technological park centre was established in Buenos Aires to improve the interaction between universities, industry and research centres in Argentina. Training was provided to young scientists.

Climate and Environment –

Study and elaboration of mathematical models for drought and flooding on a regional and global basis.

The damming effect of geopotential formations on monsoons and the ensuing droughts was demonstrated.

Protection of the coastal marine environment in the southern Mediterranean Sea.

Desertification –

Two research centres were established in China. Long-term studies and field applications showed that desertification could be forecasted; it can be stopped and even made to regress.

Energy and Pollution –

A research centre for fusion was implemented in China and training provided to young scientists of South-East Asia.

A long-term programme for the development of nanocrystalline photovoltaic energy was implemented in China, with the collaboration of the Swiss Institute of Technology.

Another long-term programme was implemented in China for the development of an energy-efficient and pollution-free carbon slurry network and research centre.

A solar research and training centre was established in Senegal for the photovoltaic applications. Training was provided to instructors and graduate students.

Extreme Weather Events –

A method for the forecasting of extreme local weather events, using simple means, was developed and made available to many developing countries. Dedicated research and operational centres were established throughout the Mediterranean Basin and training was provided to the local staff.

Floods –

A vast system for the forecasting and management of floods was implemented in the Yellow River Basin, China. The population living between the dykes, numbering in the millions, is now given warning and evacuated in case of forecasted catastrophic events.

Food –

A new source of protein for developing countries was derived from the selective cultivation of a common species of bean.

A wide variety of fruit trees was shipped to China. A research centre with laboratory and nursery was provided along with an experimental canning factory. Training was provided to the local scientists and staff both in China and in European laboratories and institutes.

Global Monitoring of the Planet –

Numerous Seminars were dedicated to the study of the search for solutions to the problems of Global Warming and the Ozone Hole.

A sophisticated seismological system was put in place in the Mediterranean Basin, with stations on all sides of the Mediterranean and linked to worldwide networks. In addition, regional seismological networks were implemented in countries prone to earthquakes.

Infectious Diseases –

A network of 15 research institutes focused their studies and field research on AIDS in 14 African countries, and two dedicated research centres were established in Uganda and Nigeria.

Medicine –

A vast programme for advanced biotechnology was undertaken in China. Training was provided in European institutes and three centres were equipped and established in Beijing and Shanghai. Two of these centres have become leading centres in China.

An epidemiological study and survey for the prevention and the treatment of heart diseases was implemented in Kenya. A specialised research and clinical centre was established and training was provided to medical personnel of various hospitals and university centres.

A vast survey of the elderly population was undertaken in China, the first of its kind. A specialised research and clinical centre was established and training provided to local medical personnel.

A long-term programme, called Kangaroo Mothers, was started in 1989 to conduct studies and seek scientific proof of the validity of a new method to care for premature babies in developing countries lacking the necessary incubators infrastructure.

Once the studies proved the method to be beneficial, a vast programme of training was implemented worldwide. The method has now even been adopted in various developed countries as it has proved to be superior to the classical incubator method.

Another pilot programme for teaching deaf-mute children was disseminated in developing countries. Training was provided to local medical personnel and specialised training centres were established in three locations. This very successful programme will help mitigate the psychological, social and economic consequences for disabled children.

Information Security –

The dedicated PMP investigated the emerging threat to the functioning of Information and Communication Technology (ICT) systems and made appropriate recommendations. The report and recommendations of the PMP on Information Security, *Toward a Universal Order of Cyberspace: Managing Threats from Cybercrime to Cyberwar* (Report), is part of an ongoing effort that has been undertaken by the WFS to address threats in this arena. The vital importance for emerging and developing countries to ensure the security of Information was stressed, to thus avoid manipulations and economic disaster. Co-operation was developed with ICT and World Summit on the Information Society (WSIS) and a joint publication issued with International Telecommunication Union (ITU).

Motivations for Terrorism –

The Permanent Monitoring Panel on the Motivations for Terrorism was established amongst others to contain and possibly eliminate the growing scourge of terrorism and its impact, to monitor research that is being conducted, organise scientific workshops to channel research efforts and reconcile conflicting views on scientific and related ethical issues, help disseminate relevant scientific data and information, develop recommendations for use by governments and international agencies, elaborate project proposals to conduct research in collaboration with scientists and others from relevant countries, and seek to operationalize the results of scientific research in the cause of deeper understanding and for the purposes of peace rather than war.

Mitigation of Terrorist Acts –

The Permanent Monitoring Panel on the Mitigation of Terrorist Acts has considered nuclear and biological megaterrorism, first to determine and to explain the magnitude of the hazard and then, in the case of biological megaterrorism, to identify the great benefit of readily-available pharmaceutical and sanitary interventions, such as hand sanitation, improvised masks, air purification, use of household bleach, in the case of a terrorist-initiated pandemic of a contagious disease such as smallpox – all to reduce the reproduction factor R of the disease from one generation of illness to the next "serial interval". Evidently, rapid development, production and use of an appropriate vaccine for the disease will minimize the serious economic and social burden of draconian public health measures.

The PMP on Terrorism Mitigation continues to maintain an overview of the capability, in principle and in practice, for the detection of the flow of fissile material that might be used by terrorists to build nuclear weapons; also the monitoring capability for the early detection and identification of disease that might be terrorist-induced or for that matter a natural outbreak, taking into account that that terrorist-induced disease might have many foci. Mitigation involves rapid assessment of the hazard, e.g., the extent and nature of radiological contamination in case of the dispersal of radioactive materials, to minimize further contamination and exposure of the population, as well as the tools and procedures to decontaminate where possible and warranted. The response involves the hard work of defining and sharing "best practices" and of sketching materials that would most effectively limit the damage from a pathogen.

The Monitoring of Planetary Emergencies –

Since 1998, sixteen Permanent Monitoring Panels (PMP) were established in order to study the fifteen classes of planetary emergencies whose total number is 72 as reported in **Figure 34**.

THE 72 PLANETARY EMERGENCIES IDENTIFIED BY THE ERICE SCIENTISTS AND CLASSIFIED INTO 15 BASIC SOURCES		NUMBER OF SOURCES
I	WATER	4
II	SOIL	3
III	FOOD	5
IV	ENERGY	5
V	POLLUTION	7
VI	LIMITS OF DEVELOPMENT	4
VII	CLIMATIC CHANGES	5
VIII	GLOBAL MONITORING OF THE PLANET	6
IX	NEW MILITARY THREATS IN THE MULTIPOLAR WORLD	3
X	SCIENCE AND TECHNOLOGY FOR DEVELOPING COUNTRIES TO AVOID A NORTH-SOUTH ENVIRONMENTAL HOLOCAUST	3
XI	THE PROBLEM OF ORGAN SUBSTITUTION	5
XII	MEDICINE, INFECTIOUS AND OTHER DISEASES	7
XIII	CULTURAL POLLUTION	8
XIV	INFORMATION SECURITY AND COMMON DEFENSE AGAINST COSMIC OBJECTS	2
XV	THE HUGE MILITARY INVESTMENTS	5
	Total	72

Figure 34

Each PMP has the aim of

- Monitoring the scientific results of research conducted in the relevant fields of science;
- Organising scientific workshops to reconcile conflicting views on scientific issues and channel research efforts;
- Helping disseminate relevant scientific data and information;
- Organising yearly group meetings and reporting on group activities to the WFS Annual General Meeting in Erice; writing recommendations for use by governments and international agencies.

Forty-six sessions of the International Seminars on Nuclear War and Planetary Emergencies have been held, since 1981, in Erice. The Proceedings are published by World Scientific (see "*A New Manhattan Project for the life and dignity of all people in the world*", Appendix 21).

A national competition was established in Italian high schools, to increase the coming generation's awareness of planetary emergencies.

A large project called EEE (Extreme Energy Events) was launched for high schools in Italy to study the origin of the cosmic rays using the most advanced detectors built at CERN by the students.

The total number of detectors will be one hundred. At the present 50% are operative all over Italy, as shown in **Figure 35**. Each star indicates the City where one detector is installed. If more than one, the number in a star corresponds to the number of detectors installed in different Institutes of the same City. The two detectors at CERN act as basic reference for the time structure of the cosmic rays showers.

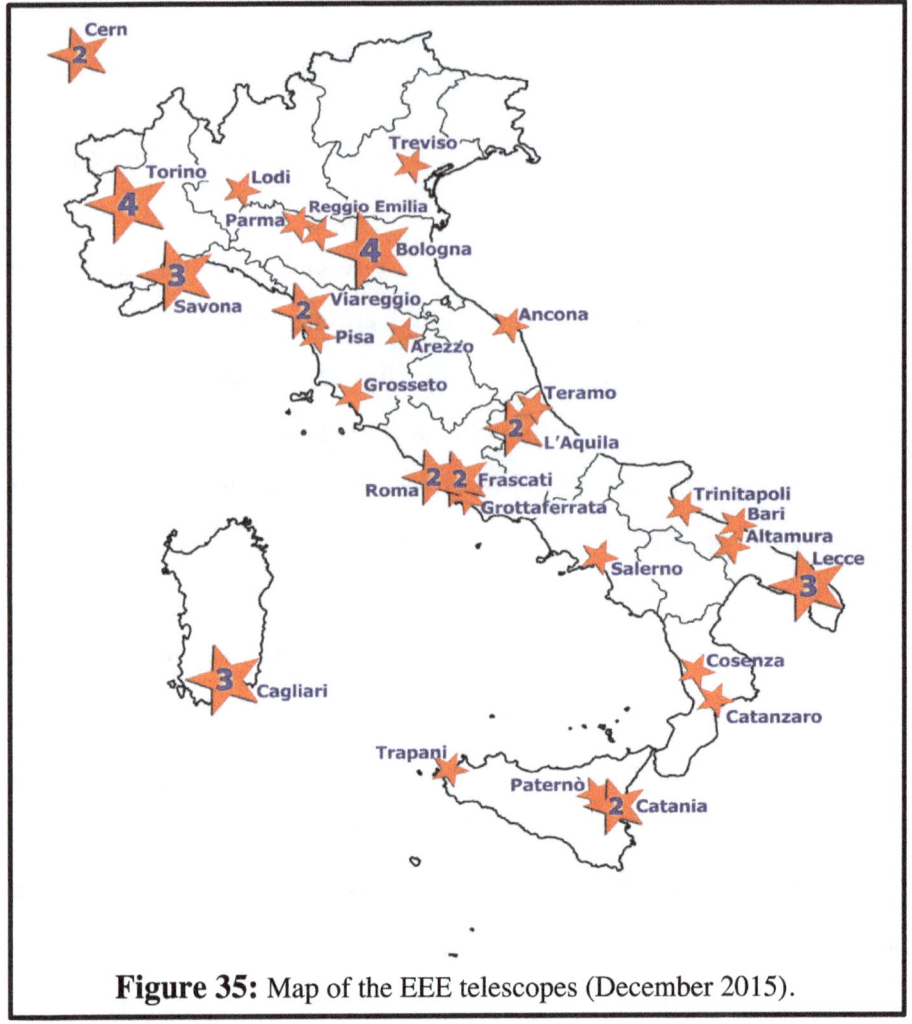

Figure 35: Map of the EEE telescopes (December 2015).

The high school students and their teachers are thus exposed to a frontier physics experiment, to which they can effectively contribute. In fact, the cosmic rays have been discovered by Victor Franz Hess more than 100 years ago. But no one has been able so far to establish where they coming from. A problem which was often recalled to us by Professor Blackett. A problem which only the Extremely Energetic Cosmic Rays can hopefully tell us sooner or later.

The purpose of the EEE Project is to bring Science outside the Ivory Towers of our Labs, as advocated by Professor Blackett.

VIII – MEMORY IS NEEDED FOR THE FUTURE

As mentioned in Chapter VI, Professor Blackett taught us young fellows intellectual modesty, recalling that Nature is smarter than all of us, as proven by the fact that

> *Fundamental discoveries*
> *in Physics*
> *are all UEEC*
> *(Unexpected Events with Enormous Consequences)*

In **Figure 36,** there is a sequence of **UEEC** events from Galilei to Blackett, Fermi-Dirac and the "strange particles", already reported in **Figure 29** when Science and History were compared.

	"UEEC" TOTALLY UNEXPECTED DISCOVERIES FROM GALILEI TO THE "STRANGE" PARTICLES
I	Galileo Galilei discovery of $F = mg$.
II	Newton discovery of $F = G \frac{m_1 \cdot m_2}{R_{12}^2}$.
III	Maxwell discovers the unification of electricity, magnetism and optical phenomena, which allows him to conclude that light is a vibration of the EM field.
IV	Planck discovery of $h \neq 0$.
V	Lorentz discovers that space and time cannot be both real.
VI	Einstein discovers the existence of time-like and space-like worlds. Only in the time-like world simultaneity does not change, with changing observer.
VII	Rutherford discovers the nucleus.
VIII	Hess discovers the cosmic rays.
IX	Dirac discovers his equation, which opens new horizons, including the existence of the antiworld.
X	Blackett discovers the "vacuum polarizations" (the first example of virtual phenomena).
XI	Fermi discovers the weak forces.
XII	Fermi–Dirac and Bose–Einstein discover two totally different statistical laws.
XIII	The "strange particles" are discovered in the Blackett Lab.
XIV	Heavy mesons with positive and negative strangeness are discovered in the Blackett Lab.

Figure 36

A few examples where I have been involved in the construction of the Standard Model are listed in **Figure 37**.

VIII – MEMORY IS NEEDED FOR THE FUTURE

---

① **The 3rd lepton, HL** (now called τ) with its own neutrino, ν_{HL} (now called ν_τ),
despite the abundance of neutrinos: ν_e and ν_μ.

② **Antimatter**
despite S-matrix and C, P, CP, T breakings.

③ **Nucleon Time-like** EM structure
despite S-matrix.

④ **No quarks** in violent (pp) collisions
despite scaling.

⑤ **Meson mixings**
$\theta_V \neq \theta_{PS} : (51°) \neq (10°) \neq 0$ *despite* $SU(3)_{uds}$.

⑥ **Effective energy:** the Gribov QCD-light
despite QCD-confinement.

⑦ **The running** of $\alpha_1\ \alpha_2\ \alpha_3$ versus **energy:**
the EGM effect and the GAP between E_{GUT} and E_{SU}.

Figure 37

The point ① in **Figure 37**, refers to the third lepton HL already discussed.

The **point** ② refers to the discovery of Antimatter [101]. Few words to explain the "despite S-matrix" and C, P, CP, T breakings.

The successes of the S-matrix theory and the crisis of the Relativistic Quantum Field Theory (RQFT) description of the fundamental interactions, in addition to the violation of the fundamental symmetry operators, gave, in the mid-sixties of last century, a central position to the search for the first example of Nuclear Antimatter [101].

The observation of the Antideuteron in 1965 was the first experimental proof that Nuclear Antimatter exists, independent of what theorists were thinking at that time. These were times of great troubles because symmetry operators had been experimentally found not to be valid in real life and no one was able to build a theory of strong interactions along the basic lines of a RQFT.

Let us try to give a synthesis of the various reasons used against RQFT, which also implies a support for S-matrix.

Lev Landau in his paper *"Fundamental Problems"*, published in "Pauli Memorial Volume" [Interscience, New York, p. 245 (1960)], wrote: *"It is well known that theoretical physics is at present almost helpless in dealing with the problem of strong interactions. ..."*, and quoting F. Dyson he concluded: *"... the correct theory will not be found in the next hundred years"*. The validity of the celebrated CPT theorem was based on RQFT but the strong forces seemed to open an entirely new horizon in physics. A horizon where there was no RQFT. *"Field theory was in disgrace, S-matrix theory was in full bloom"* recalls D. Gross in his lecture at the Third International Symposium on History of Particle Physics [Cambridge University Press (1994)].

The S-matrix theory was the antidote of RQFT. The collapse of the symmetry operators, and the fact that strong interactions seemed not to need a RQFT, focused our attention on the importance of establishing whether nuclear binding forces were CPT invariant. The success of describing strong interactions via S-matrix theory was a point of great relevance in my discussions with the CERN-DG when I had to convince him to support the construction of the "partially separated", negatively charged beam [87] and the R&D work needed for the most advanced Time-Of-Flight (TOF) electronic device to be built. To establish the existence of the Antideuteron became badly needed.

The dominant strong interaction theory of the sixties was not a RQFT but the S-matrix theory. The S-matrix theory negates completely the RQFT, upon which the CPT theorem is based. So, not only many symmetry operators (C, P, CP, T) had been found to be broken, but, at the same time, the only theory able to describe strong interactions was not a RQFT but the negation of it.

The collapse of the symmetry operators has its origin in experimental physics, with the (θ–τ) puzzle reported in **Figure 6**. The collapse of the fundamental mathematical structure, RQFT, has its origin in theoretical physics. Both structures were needed for Matter-Antimatter symmetry to hold.

After some decades of successes, RQFT started to show its severe limitations and deep troubles.

In QED, Landau discovered the "zero-charge" problem by studying the high energy behaviour of QED, concluding that the physical charge vanishes, no matter the value of the bare charge, as we let the ultraviolet cut-off become infinite. This ultraviolet limit is needed in order to achieve a Lorentz-invariant theory [L.D. Landau and I. Pomeranchuk, Dokl. Akad. Nauk SSSR 102, 489 (1955)]. Thus Landau concluded: "*We reach the conclusion that within the limits of formal electrodynamics a point interaction is equivalent, for any intensity whatever, to no interaction at all*".

On the other hand the renormalizability was the pillar of QED, but Dirac and Wigner, founding fathers of QED, were convinced that renormalization was a trick and that the physical meaning of renormalization was not truly understood, despite the basic new frontier opened by A. Petermann and E. Stueckelberg [102] implying the running with q^2 of all physical quantities. At the 1961 Solvay Conference Feynman said: "*I still hold to this belief and do not subscribe to the philosophy of renormalization*" [R. Feynman in "*The Quantum*

Theory of Fields", Proceedings of the 12th Solvay Conference, Interscience, New York (1961)].

In Weak Interactions the powerful and accurate description of beta decay processes was undermined by the fact that the theory was not renormalizable, thus losing any predictive power beyond the Born approximation.

In Strong Nuclear Forces the early successes of Yukawa field theory were confronted with severe difficulties, such as the infinities beyond the lowest order perturbation theory and the lack of any understanding of the dynamics of the Nuclear Strong Forces at the non-perturbative level. A totally unexpected fact was the rapid proliferation of strongly interacting mesons and baryons, thus depriving the nucleon and the pion fields of their privilege of being "fundamental". All the hadrons appeared to have the right of being considered as fundamental as the pion and the nucleon. Which field had to be used?

Thus RQFT, originally modelled to describe **Electrodynamics** and soon applied to the **Weak** and to the **Nuclear Forces**, and so appearing to be the natural tool for describing the dynamics of elementary particles, after some decades of success started to lose its power and its credibility as the basic mathematical formalism to describe the fundamental processes. In fact RQFT was not able to account for the explosion of experimental discoveries.

Furthermore, while RQFT was showing all its weakness, a completely different approach to describe Nuclear Interactions appeared to be very successful. This was the celebrated S-matrix theory, based on Unitarity and Analyticity, not on the Field concept. The basis of S-matrix theory is that the description of the interaction between particles should be based on analyticity as a **primary** rather than a **derived** concept.

The analytic S-matrix is supposed to give the appropriate framework in which to find a theory where no singularities are arbitrary but all are determined by general principles.

The Field concept involves a larger set of functions than those derived by the analytic continuation of the S-matrix. Unfortunately no one knew how to construct Fields purely in terms of analytic Scattering Amplitudes. Scattering Amplitudes are "on the mass shell" while fields imply extension to "off the mass shell".

An example, familiar to my work: in the early sixties I determined the existence of a strong form factor of the proton in the time-like q^2 range. **Form Factors** are not Scattering Amplitudes, nevertheless they do exist and they are due to strong interactions.

The conjectured analyticity properties of the nuclear scattering matrix is a very restricted concept, if compared with the concept of a Field.

S-matrix theory is not designed to describe experiments in which interactions between particle states do take place while momentum measurements are being performed. In other words all the physics due to virtual processes fell outside the physics described by the S-matrix theory.

It is interesting to recall the origin of the S-matrix. J.A. Wheeler (1937) and W. Heisenberg (1943) are the founders of the S-matrix theory. They pointed out a number of important advantages of S-matrix theory over conventional quantum field theory. However, Heisenberg and the other physicists working with S-matrix theory lost interest when they realized they had no way to compute interparticle forces.

It is later, with the so-called "maximal analyticity", that the S-matrix theory gained a dynamical content and became a competitor of quantum field theory. The development of the dynamical content in analyticity occurred during the late fifties and involved many names, including Gell-Mann, Goldberger, Landau, Mandelstam, Pomeranchuck, and the author of *"Nuclear Democracy"* G.F. Chew.

In 1961, at the 12th Solvay Conference devoted to "*The Quantum Theory of Fields*" Marvin Goldberger said: "*From a philosophical point of view and certainly from a practical one the S-Matrix approach at the moment seems to me by far the most attractive*" [Proceedings of the 12th Solvay Conference, Interscience, New York (1961)].

Geoffrey Chew, in 1963, wrote: "*Let me say at once that I believe the conventional association of fields with strong interacting particles to be empty. It seems to me that no aspect of strong interactions has been clarified by the field concept. Whatever success theory has achieved in this area is based on the unitarity of the analytically continued S-matrix plus symmetry principles. I do not wish to assert (as does Landau) that conventional field theory is necessarily wrong, but only that it is sterile with respect to the strong interactions and that, like an old soldier, it is destined not to die but just to fade away.*" [G. Chew, "*S-Matrix Theory*", W.A. Benjamin Inc. (1963)].

The crisis of RQFT was not a momentary crisis, in fact in 1973 the inventor of new field-theoretical entities, quarks and gluons, in his closing speech at the XVI International Conference on High Energy Physics M. Gell-Mann [Proceedings Vol. $\underline{4}$, 135 (1972)] said: "*Let us end by emphasizing our main point, that it may well be possible to construct an explicit theory of hadrons, based on quarks and some kind of glue, treated as fictitious, but with enough physical properties abstracted and applied to real hadrons to constitute a complete theory*". It is worth noting that the main father of quarks and gluons concludes by saying: "*Since the entities we start with are fictitious, there is no need for any conflict with the bootstrap or conventional dual parton point of view.*" In other words, it appeared to be not proper for a member of the very small RQFT club to use RQFT without apologies, to say the least.

The only Lorentz-invariant quantum theory where the concept of field was the primary ingredient appeared to be in trouble: "*A powerful dogma emerged-*

that field theory was fundamentally wrong, especially in its application to the strong interactions" [D. Gross, Proceedings of the Third International Symposium on History of Particle Physics, Cambridge University Press (1994)].

This short synthesis illustrates the fact that the dominant theory of particle physics in the sixties was not a RQFT but the S-matrix theory, the antidote of RQFT which is the basis of the CPT theorem.

The existence of the Antideuteron was predicted by the CPT theorem. But the very reason for CPT to be there had vanished. Why should the nuclear binding forces be CPT invariant if all known examples of fundamental interactions (**Electromagnetism**, **Nuclear Forces**, **Weak Interactions**) had serious troubles when described in terms of a RQFT? How could anyone take for granted that the Antideuteron had to exist on the basis of the existence of the Deuteron? Indeed, the Antideuteron could not really be there, thus confirming the enormous series of difficulties encountered in the description of all natural phenomena using the mathematical formalism of RQFT.

Thanks to the discovery of the non-Abelian Gauge Forces, RQFT is now back with great success. Moreover it has been discovered that the uniqueness of the S-matrix theory was a dream. There are as many S-matrices as we want, all satisfying the basic principles. In fact, any non-Abelian Gauge theory, with any Gauge group and an arbitrary number of fermions (provided that they are not too many in order to avoid the loss of asymptotic freedom) will have its S-matrix.

At the **points** ③ there is again "despite of S-matrix". Nowadays it has been forgotten that the theoretical front was – for a long time in the last century – dominated by the theory of S-matrix, which meant the abandonment of the basic principles of RQFT. Experimental results obtained in strong interactions gave rise to a strong support for this new theory. For example, the discovery of scaling at SLAC, mentioned in Chapter III-1, was interpreted as confirming the view that no RQFT could account for these findings.

Point ④ in **Figure 37** refers to quarks **despite** scaling. In fact, "scaling" discovered at SLAC suggested that pieces of protons (called partons by Feynman) behave as "free" non interacting constituents: this is, by now, the famous "asymptotic freedom". This implies that two protons colliding at high energy should break into their constituent parts, these being quarks and gluons. The (pp) collisions at ISR should have given fractionally charged particles. These were not observed [40], thus "scaling" played a double role: 1st to suggest that quarks should be experimentally found; and 2nd to be in contradiction with RQFT, confirming the view that something had to be wrong with RQFT, thus supporting S-matrix theory, since RQFT could not account for many experimental findings.

Point ⑤ refers to meson mixings; the point being why the mixings between the vector mesons θ_V was very large while the mixing between the pseudoscalar mesons θ_{PS} was very small. According to SU(3) flavour these mixing had to be both zero (see also Chapter III-2 and **Figure 11**).

Point ⑥ refers to QCD, including its "hidden side" solved by the "Effective Energy" as discussed in Chapter III-1.

Point ⑦ has been treated in Chapter III-3.

To his young fellows Professor Blackett was constantly recalling the famous **statement by Enrico Fermi**:

> *"Without Memory*
> *neither Science*
> *nor Civilisation*
> *could Exist."*

What has been reported so far is the proof of how important it is for our activity never to forget the errors. For those who could be tempted to believe that this is not valid in more recent times, there are the data in **Figure 38** where are shown the incredible mistakes made with the energy choice of the most powerful (e⁺e⁻) colliders. These data (**Figure 38**) are in my opening lecture at the AdA-INFN-EPS unveiling ceremony (**Figure 39**).

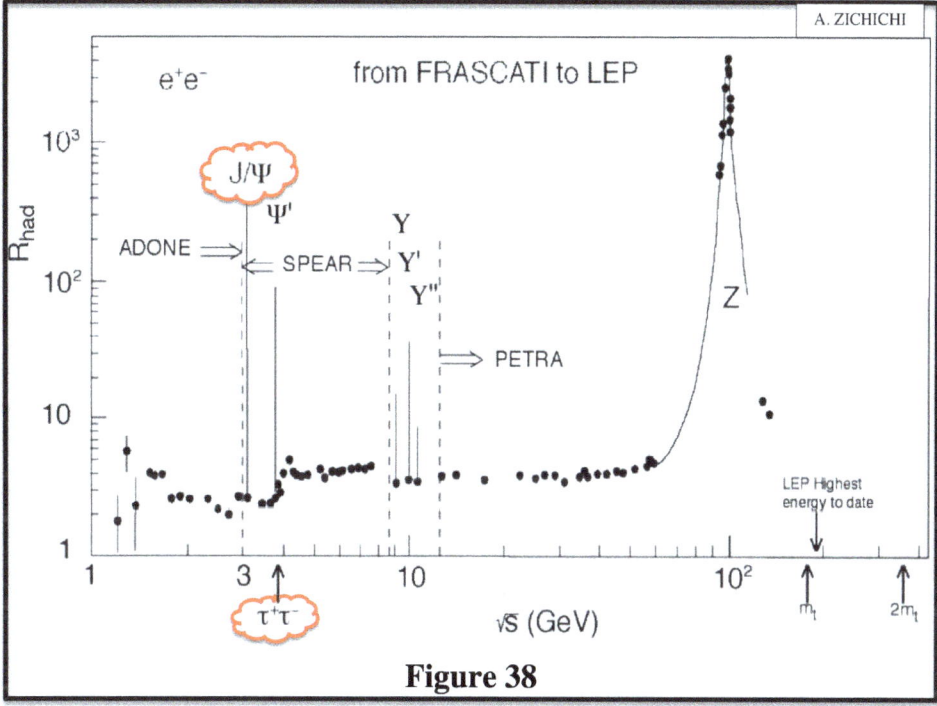

Figure 38

Few words on **Figure 38**. The first example is ADONE where the (J/ψ) could have been discovered. It would have been enough to increase the machine energy by 0.1 GeV as suggested by the CERN-Bologna group, but rejected by the people responsible. In addition to the search for "narrow resonances", the search for the third lepton (HL) proposed by the CERN-Bologna group was the reason for a slightly higher energy increase in ADONE, where the HL (now called τ) would have been discovered (as discussed in Chapters II-2, III-2 and III-3.4).

Something similar happened with PETRA, whose lowest energy was not immediately after SPEAR where the Lederman's (Y, Y', Y") would have been discovered.

These facts were reported during the special ceremony in December 2013, when the EPS gave to AdA the title of Historic Site as reported in **Figure 39**.

INFN - EPS Ceremony

European Physical Society – EPS Historic Site
The AdA Storage Ring at the INFN Frascati National Laboratories

AdA, ADONE, (J/ψ)
AND THE
3rd LEPTON

Antonino Zichichi

INFN and University of Bologna, Italy
CERN, Geneva, Switzerland
World Federation of Scientists, Beijing, Geneva, Moscow, New York

INFN Frascati National Laboratories
Thursday, December 5th, 2013 – 11 a.m.

Figure 39

VIII – MEMORY IS NEEDED FOR THE FUTURE

The great value of Blackett recalling Enrico Fermi's statement is strongly linked to his teaching about Nature being smarter than all of us, as proved by the UEEC events dominating all discoveries in Physics (**Figures 36** and **37**). Professor Blackett was eager to emphasize that **memory** is needed not only in physics but also in our other activities called civilization.

As already mentioned, in 1985, at the famous Geneva meeting when the two most powerful world leaders, Reagan and Gorbachev, met, they both agreed that the biggest enemy of peace in the world is the existence of secret labs. They both declared that these labs should be opened up, in line with the Erice Statement (**Figure 40**) signed by more than 100,000 scientists from all over the world.

ETTORE MAJORANA FOUNDATION AND CENTRE FOR SCIENTIFIC CULTURE

THE ERICE STATEMENT

- *It is unprecedented* in human history that mankind has accumulated such a military power to destroy, at once, all centres of civilization in the world and to affect some vital properties of the planet.

  The *danger* of a nuclear holocaust is not the unavoidable consequence of the great development of pure Science.

  In fact, *Science* is the study of the Fundamental Laws of Nature.

  Technology is the study of how the power of mankind can be increased.

  Technology can be for peace and for war. The choice between peace and war is not a scientific choice. It is a cultural one: the *culture of love* produces peaceful technology. The *culture of hatred* produces instruments of war. Love and hatred have existed forever. In the bronze and iron ages, notoriously pre-scientific, mankind invented and built tools for peace and instruments of war. In the so called "modern era" it is imperative that *culture of love* wins.

  An enormous number of scientists share their time between pure Science research and military applications. This is a fundamental source for the arms race.

  It is necessary that a *new trend* develops, inside the scientific community and on an international basis.

  It is of vital importance to identify the basic factors needed to start an effective process to protect human life and culture from a third and unprecedented catastrophic war. To accomplish this it is necessary to change the peace movement from a unilateral action to a truly international one involving proposals based on mutual and true understanding.

- Here are our proposals:

  1. Scientists who wish to devote all of their time, fully, to study theoretically or experimentally the basic laws of Nature, should in no case suffer for this free choice, to do only pure Science.
  2. All Governments should make every effort to reduce or eliminate restrictions on the free flow of information, ideas and people. Such restrictions add to suspicion and animosity in the world.
  3. All Governments should make every effort to reduce secrecy in the technology of defense. The practice of secrecy generates hatred and distrust. To start a ban for military secrecy will create greater stability than offered by deterrence alone.
  4. All Governments should continue their action to prevent the acquisition of nuclear weapons by additional nations or non-national groups.
  5. All Governments should make every effort to reduce their nuclear weapons stockpiles.
  6. All Governments should make every effort to reduce the causes of insecurity of non-nuclear powers.
  7. All Governments should make every effort to ban all types of nuclear tests in war technology.

- Conclusion

  Those scientists — in the East and in the West — who agree with this «Erice Statement», engage themselves morally to do everything possible in order to make the *new trend*, outlined in the present document, become effective all the world over and as soon as possible.

- This Statement was written in ERICE, August 1982, by Paul A.M. DIRAC, Piotr KAPITZA and Antonino ZICHICHI. By now the number of signatories of the Erice Statement has exceeded **100,000**, the world over.

- The 'Erice Statement' has attracted, in the eighties, the attention of World Leaders such as Deng Xiao Ping (China), Mikhail Gorbachev (USSR), Olof Palme (Sweden), Sandro Pertini (Italy), Ronald Reagan (USA), Pierre Trudeau (Canada) and stimulated various actions on their part for a Science without secrecy and without frontiers.

Figure 40

On the occasion of the twenty-fifth anniversary of the Ettore Majorana Foundation and Centre for Scientific Culture (EMFCSC), in order to promote the values of scientific culture worldwide and following a proposal by the World Federation of Scientists (WFS), a special law was voted unanimously by the Sicilian Parliament to establish the

"Ettore Majorana Prize – Erice – Science for Peace".

In my report to the Sicilian Parliament Committee it was my first point to quote what I had learned from Professor Blackett concerning the **memory**. The cover page of the special volume for the Erice Prize (**Figure 41**), recalls the Fermi statement which was so prized by our great Master.

Figure 41

VIII – Memory is needed for the Future

> The Erice Prize has been awarded to fellows who played a leading role in promoting and implementing the goals outlined in the "Erice Statement" for a Science without secrets and without borders:
>
> P.A.M. Dirac, P.L. Kapitza, A.D. Sakharov, E. Teller, V.F. Weisskopf, J.B.G. Dausset, S.D. Drell, M. Gell-Mann, H.W. Kendall, L.C. Pauling, A. Salam, C. Villi, R. Doll, J.C. Eccles, T.D. Lee, L. Montagnier, Qian Jiadong, J.S. Schwinger, U. Veronesi, G.M.C. Duby, R.L. Garwin, S.L. Glashow, D.C. Hodgkin, R.Z. Sagdeev, K.M.B. Siegbahn, Y.P. Velikhov, J. Karle, J.-M.P. Lehn, A. Magnéli, N.F. Ramsey, H. Rieben, J.J. van Rood, C.S. Wu, R.L. Mössbauer, A. Müller, H. Kohl, M.S. Gorbachev, H.H. John Paul II, R. Clark, M. Cosandey, A. Petermann, R. Wilson, Lord J. Alderdice, J.I. Friedman, M. Koshiba, S. Coleman, A.N. Chilingarov, P.C.W. Chu, L. Esaki, W.N. Lipscomb Jr., J. Szyszko, M.-K. Wu, H.A. Hauptman, D.H. Hubel, R. Huber, B.I. Samuelsson, H. Sun, A.E. Yonath, G. 't Hooft, Y.T. Lee, W. Arber and S.C.C. Ting.

In 1994 a Special Ceremony in memory of Lord Blackett took place in Erice. The participants to the Special Ceremony, great admirers of Patrick Blackett, are in the **Photos 12** and **13**.

Photo 12

LAUREATES OF THE ERICE PRIZE SCIENCE FOR PEACE FOR A SPECIAL CEREMONY IN HONOUR OF PATRICK M.S. BLACKETT

Blackett Institute, 6th November 1994: (1) Dausset, (2) Garwin, (3) Van Rood, (4) Müller, (5) Qian Jadong, (6) Velikhov, (7) Drell, (8) Glashow, (9) Sagdeev, (10) Ramsey, (11) Lee, (12) Doll, (13) Gell-Mann, (14) Siegbahn, (15) Wu, (16) Magnéli, (17) Lehn, (18) Duby, (19) Karle, (20) Veronesi, (21) Poma (Mayor of Erice), (22) Spitaleri, (23) Pauling, (24) Schwinger, (25) Salam, (26) Villi, (27) Montagnier, (28) Rieben and (29) A.Z..

Photo 13

IX – THE FUTURE

And now let us move to the Future. Here the novelty is "complexity". Once again this is the proof that Nature is smarter than all of us.

In fact, the Logic of Nature allows the existence of a large variety of structures with their regularities and laws which appear to be independent from the basic constituents and fundamental laws of Nature which govern their interactions.

But without these laws, it would be impossible to have the real world which is in front of us and which we are part of. A series of complex systems is shown in **Figure 42**.

As you can see, we go from traffic flux, to the internet network, to earthquakes and seismicity, to social and economic systems, to the behaviour of financial markets, to the study of cosmological structures, and so on.

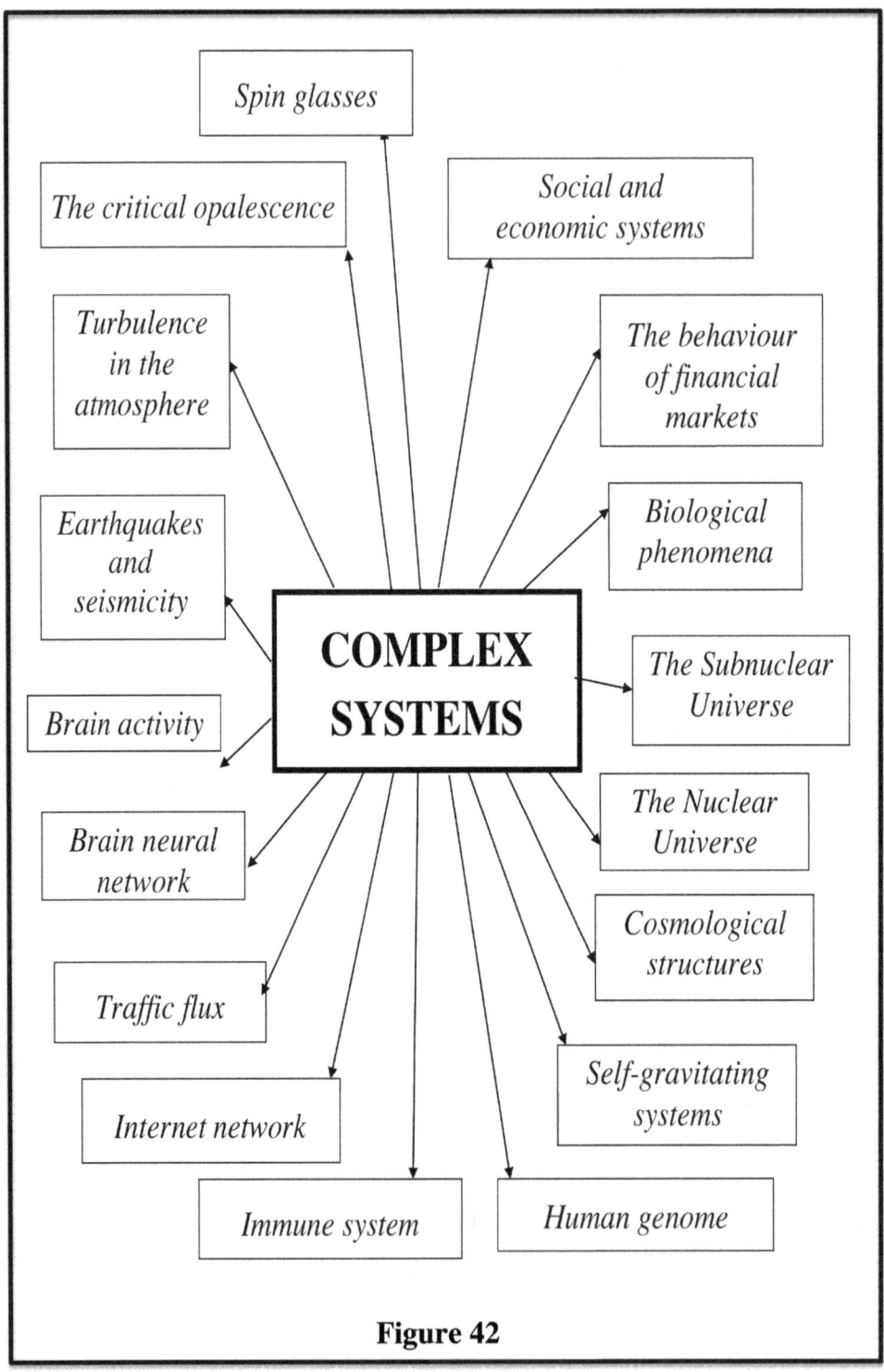

Figure 42

There are many other systems in addition to those shown in **Figure 42**.

It turns out that complexity exists at all scales including the Subnuclear Universe.

The experimental evidences for the **existence** of **complexity are two**: the **AFB** phenomena and the **UEEC** effects.

1) **The Anderson-Feynman-Beethoven-type phenomena (AFB)** i.e. phenomena whose laws and regularities ignore the existence of the Fundamental Laws of Nature from which they originate. In **Figure 43** there are four examples. Starting with Beethoven and the laws of acoustics and going on to the study of matter endowed with the privilege of having on it the structure and the properties which allow "life" to be there. As far as we can go, all we can say is that all known fundamental processes which are needed for life are due to the electromagnetic forces. At a more fundamental level we know that the nuclear forces are not fundamental. In other words they are secondary effects whose roots are in Quantum ChromoDynamics. The fourth example refers to the Superworld which needs a Space-Time with 43 dimensions.

AFB PHENOMENA FROM BEETHOVEN TO THE SUPERWORLD

Beethoven and the laws of acoustics.

Beethoven could compose superb masterpieces of music without any knowledge of the laws governing acoustic phenomena.

But these masterpieces could not exist if the laws of acoustics were not there.

The living cell and QED.

To study the mechanisms governing a living cell, we do not need to know the laws of electromagnetic phenomena whose advanced formulation is QED (Quantum ElectroDynamics).

All mechanisms needed for life are, to a great extent, examples of electromagnetic processes. If QED was not there, Life could not exist.

Nuclear Physics and QCD.

Proton and neutron interactions appear as if a fundamental force of nature is at work: the nuclear force, with its rules and its regularities.

These interactions ignore that protons and neutrons are made of quarks and gluons.

Nuclear Physics does not appear to care about the existence of Quantum ChromoDynamics (QCD), the fundamental force acting between quarks and gluons at the heart of the subnuclear world.

Nuclear Physics ignores QCD but all phenomena occurring in Nuclear Physics have their roots in the interactions of quarks and gluons.

In other words, protons and neutrons behave like Beethoven: they interact and build up Nuclear Physics without 'knowing' the laws governing QCD.

The most recent example of AFB type phenomenon comes from the frontier of our scientific knowledge: the Superworld. What we call the **World** could apparently not care less about the existence of the **Superworld**, whose foundation is not in the four dimensional Space-Time but in a Superspace with 43 dimensions.

Figure 43

2) **The Sarajevo-type effects, i.e. Unexpected Events of quasi irrelevant magnitude which produce Enormous Consequences (UEEC)** (some examples are in **Figures 36** and **37**).

The only certainty about complexity is the existence of these experimentally observable effects. Nature tells us that UEEC and AFB exist at all scales, and therefore complexity exists at all scales, as illustrated in **Figure 44**. How Nature goes from the lowest degree of Complexity (Science) to the highest degree of Complexity (History) is in **Figure 30** (Chapter VI).

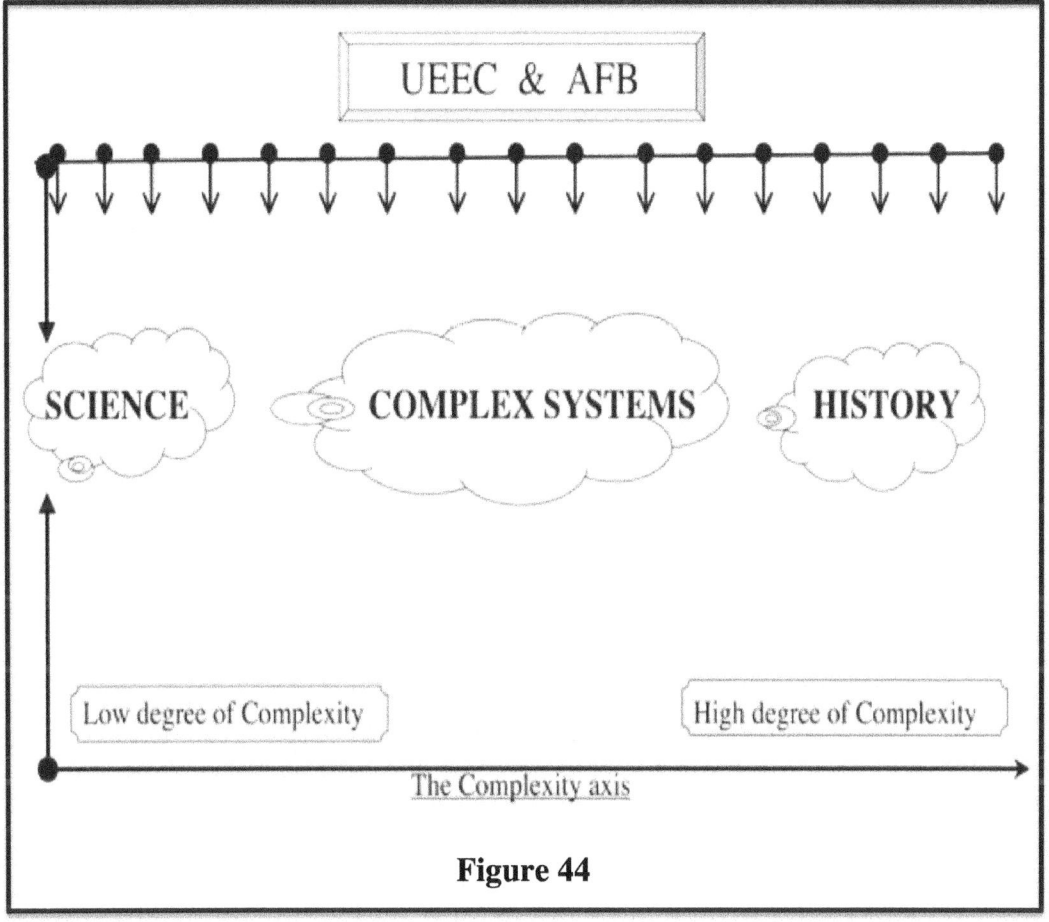

Figure 44

Details on Subnuclear Physics, the Superworld, the ELN and other arguments mentioned later in Chapter X, can be found in my book [103] whose front page is reproduced below.

World Scientific Series in 20th Century Physics – Vol. 24

Subnuclear Physics

The First 50 Years: Highlights from Erice to ELN

Antonino Zichichi

Edited by
O. Barnabei, P. Pupillo & F. Roversi Monaco

World Scientific

THE LESSON NEEDED FOR THE FUTURE

For the Future what is to be Expected?

We have proved that AFB and UEEC – which are at the origin of Complexity, with its consequences permeating all our existence, from molecular biology to life in all its innumerable forms up to our own, including History – do exist at the fundamental level [104–108].

It turns out that Complexity in the real world exists, no matter the mass-energy and Space-Time scales considered.

Therefore the only possible prediction is that:

- **Totally Unexpected Effects** should **show up**.
- **Effects**, which are impossible to be predicted on the basis of **present knowledge**.

We should be prepared with **powerful experimental instruments**, **technologically at the frontier of our knowledge**, to discover Totally Unexpected Events in all laboratories the world over (including CERN in Europe, Gran Sasso in Italy, and other facilities in Japan, USA, China and Russia). All the pieces of our Physics could not have been discovered if the experimental technology was not at the frontier of our knowledge.

Examples include: the cloud-chambers (**Blackett**, Anderson, Neddermeyer), the photographic emulsions (Lattes, Muirhead, Occhialini, Powell), the high power magnetic fields (Conversi, Pancini, Piccioni) and the powerful particle accelerators and associated detectors for the discovery – the world over – of the **SM&B** (see later Chapter X and **Figure 48**).

This means that we must be prepared with the most advanced technology for the discovery of totally unexpected events like the ones found in the **Blackett group**, from (e^+e^-) production to strange particles.

The mathematical descriptions, and therefore the predictions – for new phenomena to be discovered in the field opened by the given UEEC event – come after the UEEC event, never before.

Recall:

• The **discoveries in Electricity**, **Magnetism** and **Optics** (UEEC).

• **Radioactivity** (UEEC).

• The **Cosmic Rays** (UEEC).

• The **Weak Forces** (UEEC).

• The **Nuclear Physics** (UEEC).

• The **Strange Particles** (UEEC).

• The **3 Columns** (UEEC).

• The **origin of the Fundamental Forces** (UEEC).

X – CONCLUSIONS – FROM BLACKETT TO PRESENT DAY PHYSICS

From 1947 to the present, a large amount of knowledge has been gained. This is expressed in terms of three fundamental forces (**Figure 45**) and three columns (**Figure 46**). They are needed in order to build the Standard Model (**Figure 47**) and Beyond, as reported in (**Figure 48**).

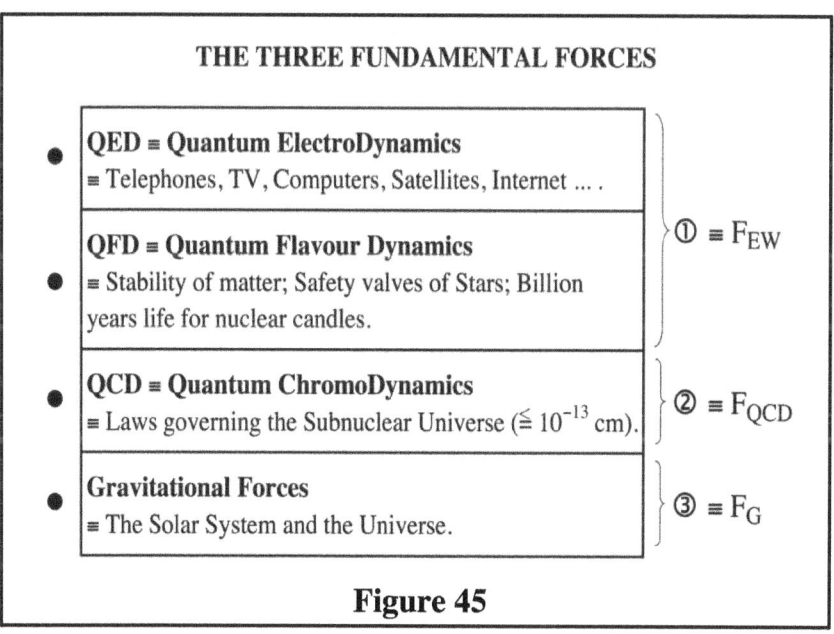

Figure 45

149

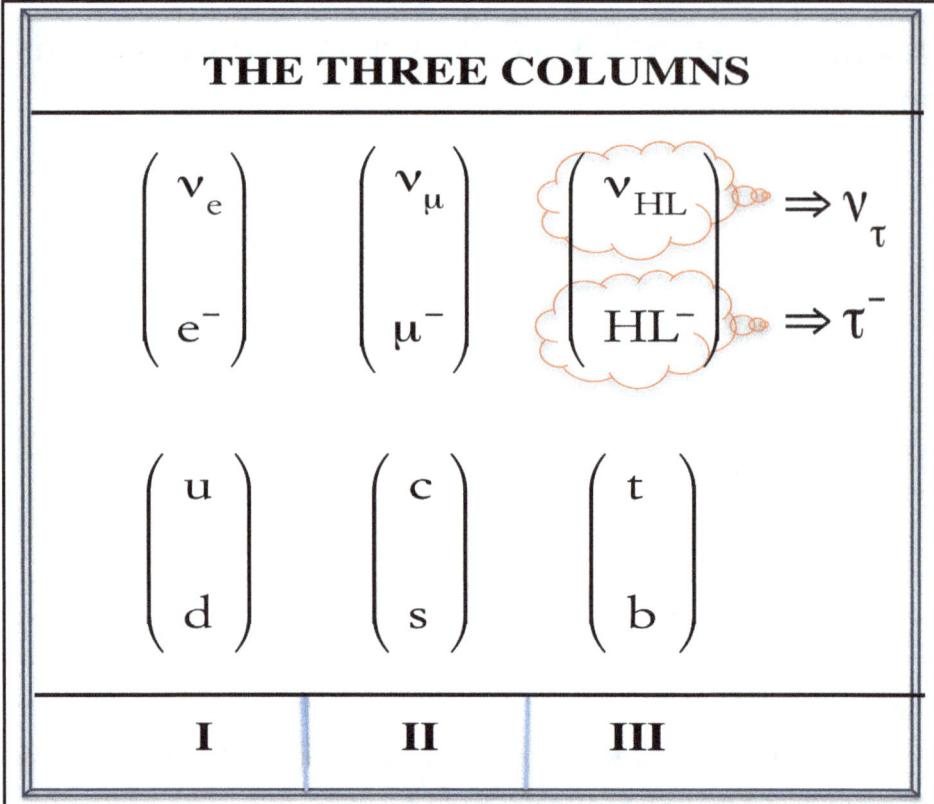

Figure 46 (see "*The Useless Particles and the Origin of the Universe*", Appendix 22).

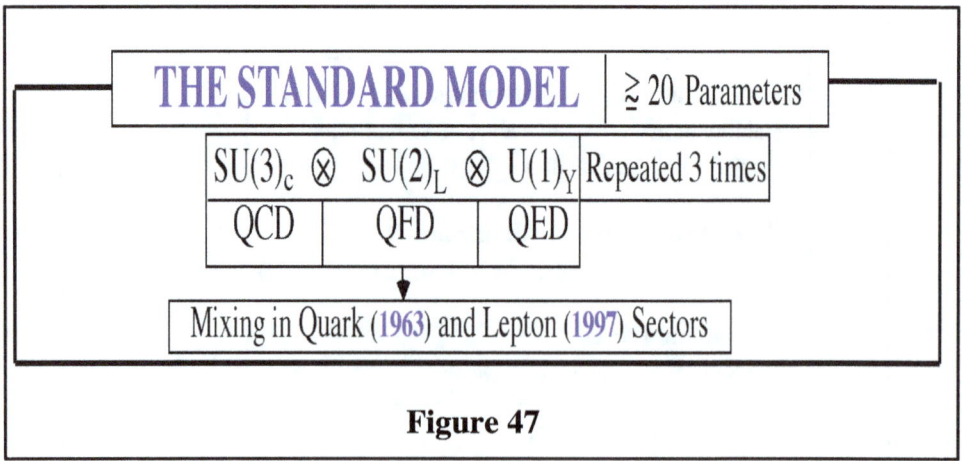

Figure 47

We are sure that it is necessary to go beyond the Standard Model whose five basic steps are reported in **Figure 48**.

SM&B
THE STANDARD MODEL AND BEYOND
THE FIVE BASIC STEPS NEEDED FOR SM&B

① The renormalization group equations (RGEs) imply that the gauge couplings (α_i) and the masses (m_j) all run with k^2. It is this running which allows GUT, suggests SUSY and produces the need for a non point-like description (RQST) of Physics processes, thus opening the way to quantize gravity.

RGEs (α_i (i ≡ 1, 2, 3); m_j (j ≡ q, l, G, H)) : $f(k^2)$.
- GUT ($\alpha_{GUT} \cong 1/24$) & GAP ($10^{16} - 10^{18}$) GeV.
- SUSY (to stabilize $m_F/m_P \cong 10^{-17}$).
- RQST (to quantize Gravity).

② All forces originate in the same way: the gauge principle.
Gauge Principle (hidden and expanded dimensions).
— How a Fundamental Force is generated: SU(3); SU(2); U(1) and Gravity.

③ Imaginary masses play a central role in describing nature: SSB & Confinement.
The Physics of Imaginary Masses: SSB.
— The Imaginary Mass in SU(2)×U(1) produces masses thanks to SSB: ($m_{W^\pm}$; m_{Z^0}; m_q; m_l), including $m_\gamma = 0$.
— The Imaginary Mass in SU(5)⇒SU(3)×SU(2)×U(1) or in any higher (not containing U(1)) Symmetry Group ⇒ SU(3)× SU(2)×U(1) produces Monopoles.
— The Imaginary Mass in SU(3)$_c$ generates Confinement.

④ The mass-eigenstates are mixed when the Fermi forces come in: the matrix describing the mixing is the product of two fundamental matrices.
Flavour Mixings & CP ≠ , T ≠ (direct ≠ , not via SSB).
— Why is the mixing there? No need for it but it is there.

⑤ The Abelian force QED has lost its role of being the guide for all fundamental forces. The non-Abelian gauge forces dominate and have features which are not present in QED.
Anomalies & Instantons.
— Basic Features of all Non-Abelian Forces.

Figure 48

The note below is needed for the symbols used in **Figure 48**.

	NOTE	
q	≡	quark and squark;
l	≡	lepton and slepton;
G	≡	Gauge boson and Gaugino;
H	≡	Higgs and Shiggs;
m_F	≡	Fermi mass scale;
m_P	≡	Planck mass scale;
k	≡	quadrimomentum;
C	≡	Charge Conjugation;
P	≡	Parity;
T	≡	Time Reversal;
≠	≡	Breakdown of Symmetry Operators;
RGEs	≡	Renormalization Group Equations;
GUT	≡	Grand Unified Theory;
SUSY	≡	Supersymmetry;
RQST	≡	Relativistic Quantum String Theory;
SSB	≡	Spontaneous Symmetry Breaking.

Where can SM&B come from? The answer to this question needs the most powerful machine in the world mentioned in Chapter VII, the ELN. A picture of the model is shown in **Figure 49**.

SM&B TELLS US THAT ELN IS NEEDED

Figure 49: A view of the 300 km ring.

ELN could be built with an extrapolation of the presently known technologies for a total energy of 400 TeV and a peak luminosity of

$$\cong 10^{35} \cdot cm^{-2} \cdot sec^{-1}.$$

The maximum energy could be 1 PeV but, in order to reach this energy and the peak luminosity quoted above R&D are needed.

The ELN project is a very ambitious one but we should be encouraged by our previous experiences. In fact, the path leading to the ELN project has gone through the Gran Sasso project, now the largest and most powerful underground laboratory in the world, the LEP-white-book, which allowed this great European venture to overcome the many difficulties that had prevented its implementation during many years, the HERA collider, now successfully completed, the roots of LHC as, for example, the 27 km long (rather than the 13 km initially planned) LEP tunnel whose dimensions are big enough for two colliders (not only for the (e^+e^-) initially planned) and the LAA-R&D project, implemented to study the detector technologies needed for the LHC and future colliders.

These past achievements in project realization are mentioned in order to corroborate my optimism and enthusiasm in encouraging new actions and new ideas for the **future of Subnuclear Physics in Europe and the world over**, all having CERN as the focal point, the greatest Subnuclear Physics Lab in the world.

The history of mankind in terms of its achievements, as reported in **Figure 50**, is the source of many questions.

The biggest one being: "Why were 1800 years lost - from Archimedes to Galilei?" A question pointed out by Galilei himself.

X – CONCLUSIONS – FROM BLACKETT TO PRESENT DAY PHYSICS

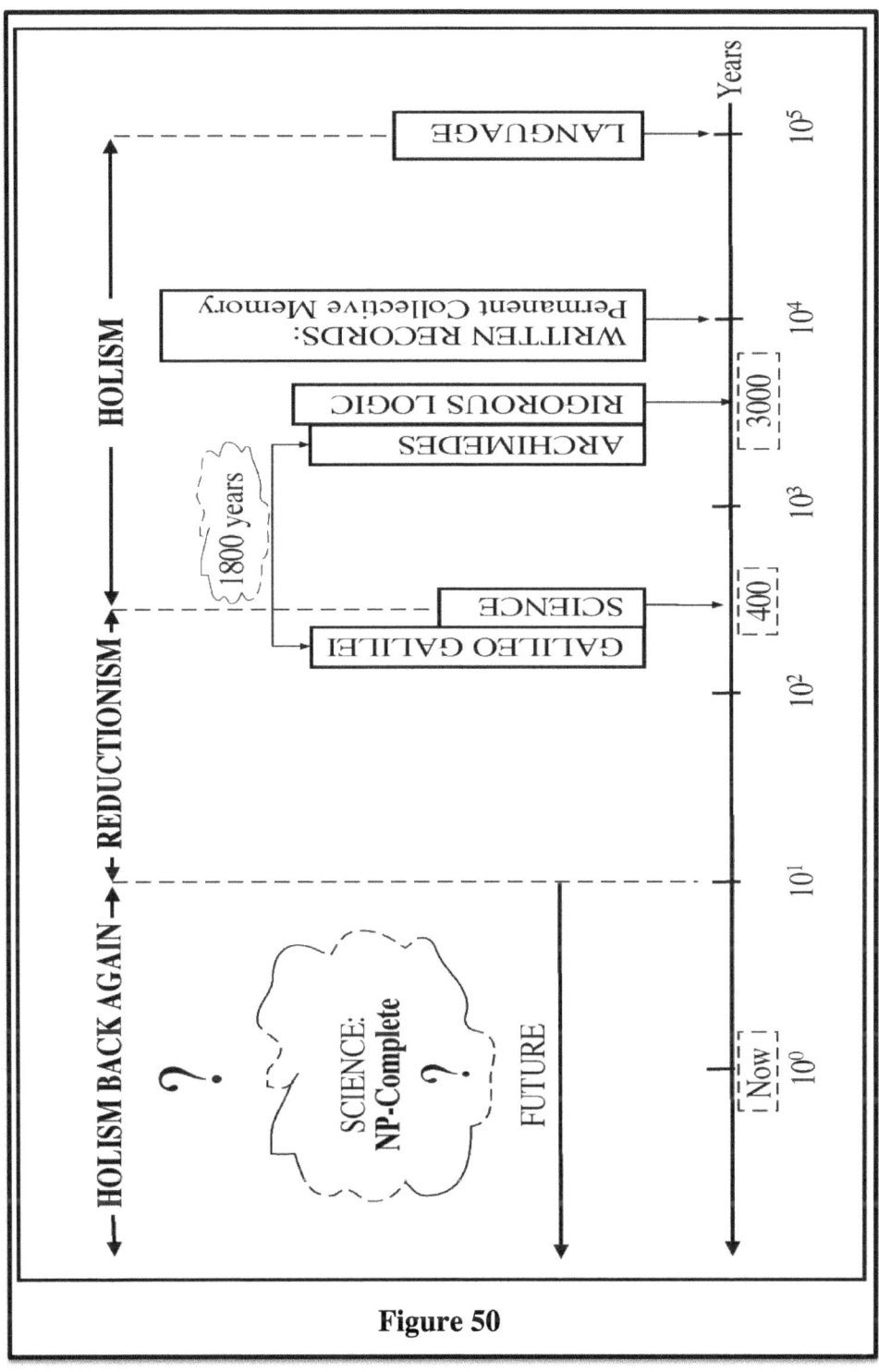

Figure 50

All this started with the invention of Language (10^5 years), going through the invention of Permanent Collective Memory (written Language), and finally the discovery of Rigorous Logic with Archimedes (c. 287 BC – c. 212 BC).

Here are the words that Marcus Tullius Cicero (107 BC – 43 BC) wrote in his *De Republica* when one of his characters, Philus, describes what the Roman General Marcus Marcellus did when his troops ran to conquer the City of Syracuse. The General took only one thing for himself: Archimedes' mechanical sphere. The Philus saw how the sphere was working and said: *"The famous Sicilian has been endowed with greater genius than one would imagine it possible for a human being to possess."*

Why was it that another 1800 years were needed for the discovery of Galileian Science?

Following the optimism of Professor Blackett there is no limit to our knowledge, provided we do not forget that the Logic of Nature can only be deciphered by rigorous, reproducible experimental proofs.

It could be that science will be mathematically proven to be "NP-complete". This is the big question for the immediate future [109].

It is therefore instructive to recall what I learned when I joined the Blackett group: how science fits into the whole of our knowledge as reported in **Figure 31**, here reproduced (**Figure 31-bis**) to help the reader.

Nowadays one of the biggest problems lies in understanding how **the Logic of Nature** allows us to go **from the lowest limit of Complexity, Science, to the highest limit of Complexity, History**. This is exactly the way Professor Blackett was focusing his interests in his attempt to understand the Logic of Nature: **Science** and **History** being the two limits of complex systems as shown in **Figure 30** (Chapter VI).

From 1974 a series of Scholarships and Diplomas to honour his memory have been awarded to students of the Subnuclear Physics Schools, as reported in **Figure 51**.

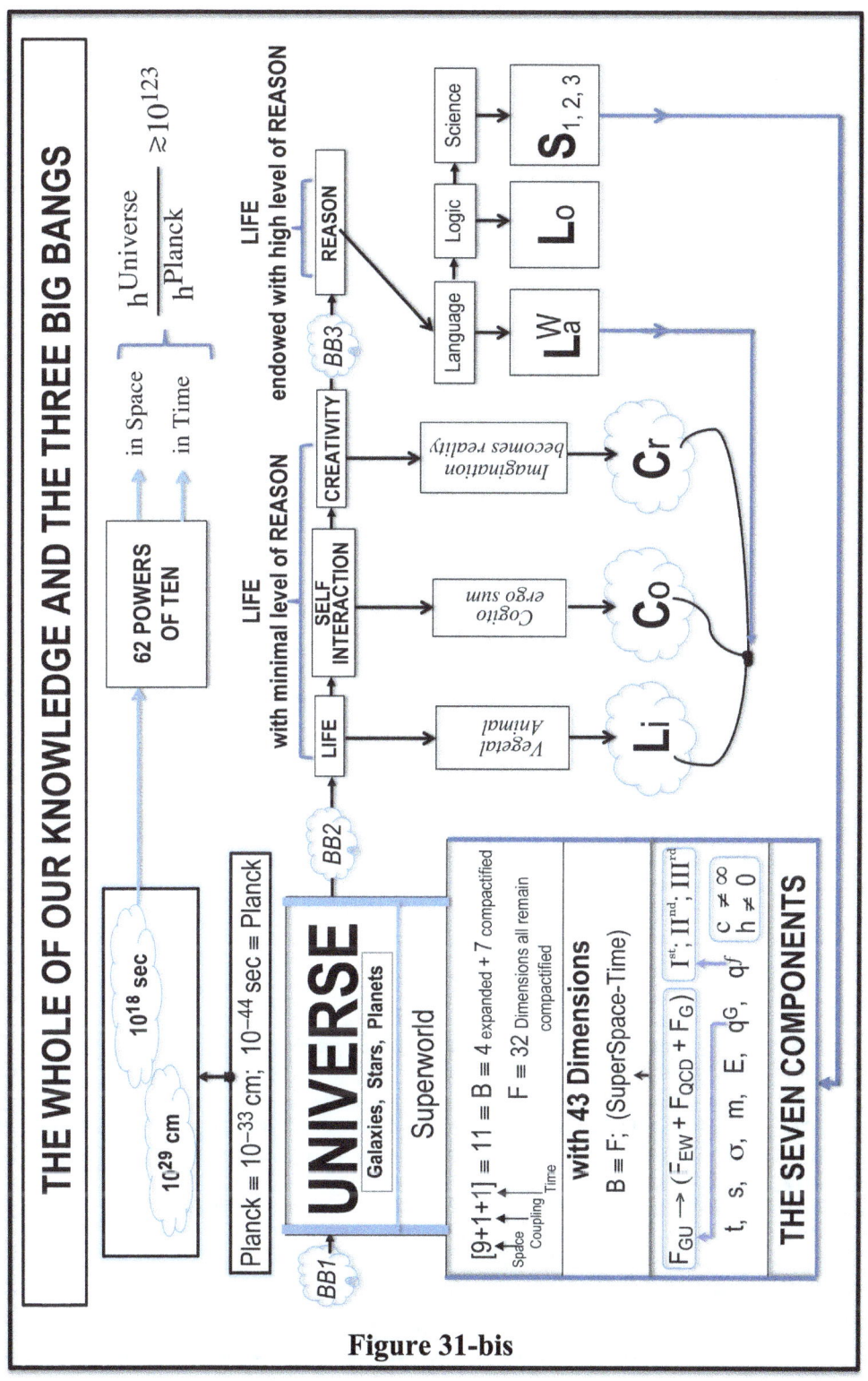

Figure 31-bis

«ETTORE MAJORANA» FOUNDATION AND CENTRE FOR SCIENTIFIC CULTURE
INTERNATIONAL SCHOOL OF SUBNUCLEAR PHYSICS
Patrick M.S. Blackett
Scholarships and Diplomas from 1974

1974 Jonathan Schonfeld, *Princeton University, Princeton, NJ, USA*
1975 Michael E. Peskin, *Cornell University, Ithaca, NY, USA*
1976 William J. Marciano, *Rockefeller University, New York, USA*
1977 William Celmaster, *Harvard University, Cambridge, MA, USA*
1978 Anthony D. Kennedy, *University of Sussex, Brighton, Sussex, UK*
1979 Paul Ginsparg, *Cornell University, Ithaca, NY, USA*
1980 Enore Guadagnini, *University of Pisa, Italy*
1981 Jay L. Banks, *Ohio State University, Columbus, OH, USA*
1982 Leonardo Castellani, *State University of New York, Stony Brook, NY, USA*
1983 Robert H. Bernstein, *University of Chicago, IL, USA*
1984 Rosario Nania, *CERN, Geneva, Switzerland*
1985 Gerard Gilbert, *The University of Texas, TX, USA*
1986 Andrea Pasquinucci, *University of Milano, Italy*
1987 Jun Liu, *University of Texas at Austin, TX, USA*
1988 Gerjan Bobbink, *NIKHEF, Amsterdam, The Netherlands*
1989 Bindu A. Bambah, *Panjab University, Chandigarh, India*
1990 Mark Wexler, *Princeton University, Princeton, NJ, USA*
1991 Matthias Neubert, *Universität Heidelberg, Heidelberg, Germany*
1992 Rongzhi Liu, *Caltech, Pasadena, CA, USA*
1993 Konstantinos Skenderis, *State University of New York, Stony Brook, NY, USA*
1994 Nicola Fabiano, *INFN-Frascati, Italy*
1995 Vladimir Anferov, *University of Michigan, Ann Arbor, MI, USA*
1996 Stefan Schönert, *Technische Universität, München, Germany*
1997 William E. Brown, *University of Oxford, UK*
1998 Salvatore Mele, *CERN, Geneva, Switzerland*
1999 Alan Marcus, *Tel Aviv University, Israel*
2000 Arno Straessner, *CERN, Geneva, Switzerland*
2001 Kristian Harder, *DESY, Hamburg, Germany*
2002 Mark Ramtohul, *University of Edinburgh, UK*
2003 Wolfgang Menges, *DESY, Hamburg, Germany*
2004 Sevil Salur, *Yale University, New Haven, CT, USA*
2005 Kazunori Hanagaki, *Fermilab, Batavia, IL, USA*
2006 Paolo Aschieri, *University of Piemonte Orientale, Alessandria, Italy*
2007 Mira Krämer, *DESY, Hamburg, Germany*
2008 Zachary Marshall, *Columbia University, New York, NY, USA*
2009 Mickey Guotai Chiu, *Brookhaven National Laboratory (BNL), Upton, NY, USA*
2010 Mohd Danish Azmi, *Aligarh Muslim University, Aligarh, India*
2011 Magnus Mager, *CERN, Geneva, Switzerland*
2012 Max Zoller, *Karlsruhe Institute für Tecnologie, Karlsruhe, Germany*
2013 Silvia Nagy, *Imperial College, London, UK*
2014 Teresa Lenz, *Hamburg University, Germany*
2015 Alexandros Anastasiou, *Imperial College, London, UK*

Figure 51

XI – THE *"PIERSANTI MATTARELLA TOWER OF THOUGHT"* AND THE VIEW WHICH ENCHANTED PROFESSOR BLACKETT

When Blackett visited Erice,
what would become the

"Piersanti Mattarella Tower of Thought"

was only
the highest point of sight in Erice.
From this highest point in Erice (**Photo 14**)
you can see
the view (**Photo 15**)
that enchanted Professor Blackett
and since 1963 continues to enchant all fellows
(two examples in **Photos 16 and 17**)
participating in the activities
of the 126 International Schools of the EMFCSC.

Photo 14: The "**Piersanti Mattarella Tower of Thought**" in the Rabi Institute.

The highest point of sight in Erice was inside a convent of nuns, which is now the Isidor Isaac Rabi Institute of the EMFCSC.

XI – THE "PIERSANTI MATTARELLA TOWER OF THOUGHT" AND THE VIEW WHICH ENCHANTED PROFESSOR BLACKETT

Photo 15

Photo 16: From left Sheldon Glashow, Peter Higgs and Gerardus 't Hooft (Erice 1997). Glashow was the first to identify the correct group for the electroweak interactions: SU(2)×U(1). Higgs was the first to have the correct idea on how to introduce masses without producing disasters. 't Hooft was the only fellow who knew how to prove that the non-Abelian electroweak interactions described by the gauge forces SU(2)×U(1) are renormalizable. Furthermore he was the first to discover that the β–function has negative sign (asymptotic freedom of QCD).

Photo 17: From left Graham M. Shore, Ludvig D. Faddeev, Yakov Azimov, Masatoshi Koshiba, A.Z., Edward Witten, Bjorn H. Wiik, Sheldon L. Glashow and Gerardus 't Hooft at Erice (1998).

XI – THE "PIERSANTI MATTARELLA TOWER OF THOUGHT" AND THE VIEW WHICH ENCHANTED PROFESSOR BLACKETT

This **"Tower of Thought"** was blessed in 1993 by H.H. John Paul II (**Photo 18**).

Photo 18: The Pope H.H. John Paul II after the blessing of the
"**Piersanti Mattarella Tower of Thought**".
Behind A.Z. is Yuri A. Izrael, the scientist who managed to avoid rain on the Chernobyl reactor in the course of several weeks.

At this highest point in Erice, John Bell had been thinking about the Bell's inequality and two other friends, Rudolf L. Mössbauer e K. Alex Müller, told me they were inspired in the right direction for the Nobel Prize, while working in the Tower.

What would later be the Blackett Institute was a badly collapsed monument: the San Domenico, once the most famous Church in Erice. In 1963 only the main entrance of the Church plus its four basic walls existed. The monument has been reconstructed – thanks to a special law established by the President of the Sicilian Government Piersanti Mattarella. This is how, using the most advanced technology, the Church became the main lecture hall of the centre named after Dirac. On its top, the "Olof Palme Discussion Hall" (**Photo 19**) could be constructed.

Photo 19: The "**Olof Palme Discussion Hall**" in the Blackett Institute.

XI – THE "PIERSANTI MATTARELLA TOWER OF THOUGHT" AND THE VIEW WHICH ENCHANTED PROFESSOR BLACKETT

The activities of the rising "Ettore Majorana" Subnuclear Physics School in Erice started to be implemented in 1963. The first Subnuclear Physics School took place in 1963 and the father of Piersanti, Bernardo Mattarella, who was Minister of the Italian Government came to attend the Opening Session (**Photo 20**).

The Opening Session
of the 1st Subnuclear Physics School
(26 May 1963).
From left: Professor Victor Weisskopf,
H.E. Rt. Hon. Bernardo Mattarella,
Minister of the Italian Government,
A.Z. and Professor Sidney Drell.

Photo 20

The Minister Bernardo Mattarella, also came to the Opening Ceremonies of the second (1964) and the third (1965) Subnuclear Physics Schools. In the second two discoveries, Ω^- and $K_2^0 \to \pi^+ \pi^-$, were presented. The Figure below reproduces the foreword and the cover page of the volume, where all lectures and discussions are reported.

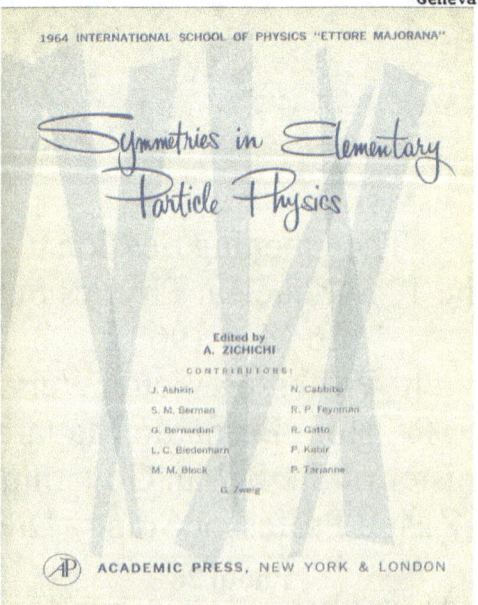

XI – THE "PIERSANTI MATTARELLA TOWER OF THOUGHT" AND THE VIEW WHICH ENCHANTED PROFESSOR BLACKETT

The son of Bernardo Mattarella, Piersanti, became President of the Government of Sicily in 1978. It is thanks to his strong support, as said before, that it was possible to transform the old and collapsed San Domenico Church into the Blackett Institute.

> The scientific community of the Erice Schools
> wanted the
>
> **Tower of Thought**
>
> to be named
>
> **Piersanti Mattarella**
>
> when in 1980
> the President of the Sicilian Government
> was assassinated.
>
> * * * * * * * *
>
> A special ceremony was dedicated
> to the memory of Piersanti Mattarella in 1988.
> The Minister of the Italian Government
> **Sergio Mattarella**
> came to the Centre
> where many scientists,
> the CERN Director General, Herwig Schopper,
> many Nobel Laureates (Salam, T.D. Lee, Siegbahn)
> and other distinguished physicists
> contributed to the ceremony (**Photos 21** and **22**)
> including the Scientific Advisor of President
> Deng Xiao Ping (Qian Jiadong).

Photo 21: From left: Minko Balkanski, Abdus Salam, Kai M.B. Siegbahn, the Minister for Relations with Parliament Sergio Mattarella, A.Z., Tsung-Dao Lee, Herwig F. Schopper and Qian Jiadong.

Photo 22: From left: Enzo Boschi, Minko Balkanski, Abdus Salam, Kai M.B. Siegbahn, the Minister for Relations with Parliament Sergio Mattarella, A.Z., Tsung-Dao Lee, Herwig F. Schopper, Qian Jiadong, Arthur S. Wightman, Günther Harbeke and Arthur I. Miller.

*Fifty years later
– in 2013 –
in the Dirac Lecture Hall
of the Blackett Institute
a
fundamental problem
of Subnuclear Physics
was dedicated (**Photo 23**) to the memory of*

Piersanti Mattarella

and

Olof Palme,

*two strong supporters of the
Ettore Majorana Centre in Erice.*

Photo 23: The **"Paul A.M. Dirac" Lecture Hall** in the Blackett Institute during the International School of Subnuclear Physics 2013 where a fundamental problem was dedicated to Piersanti Mattarella and Olof Palme.

XII – HOW THINGS REALLY HAPPEN

Certainly **not** as it is often reported and believed to be true. Let me give a few examples.

① Very few people know how it happened that the UK Government – who was not willing to support the Rabi's proposal at the Unesco 1950 Conference in Florence to create a European Laboratory, CERN – changed its position and decided to support the establishment in 1954 of CERN in Geneva.

I was often in Professor Blackett's office. The topics were the new frontiers of Physics, dominated by the existence of a new fundamental charge of Nature, called "strangeness". As we have seen these new frontiers were all coming from the discovery of the very strange events, named by Blackett "V–particles". Fermi's criticism had to be solved. If it is really true that a new "**charge**" exists, why it is so difficult to find the proof so much wanted by Enrico Fermi? Why in the cosmic rays had no one observed the associated production of a pair of "heavy mesons" (called θ–particles at the time) with positive and negative "strange" charge? In the USA a new accelerator was working at the high energy needed to produce pairs of these mesons. But no evidence had been observed for the pair production of strange mesons. After all we should not forget that the decisive proof concerning the production of electric charges was the experimental discovery by Blackett and Occhialini, in 1932, of the pair production of an electron and an antielectron, as theoretically predicted by Dirac. This was the starting point of Quantum ElectroDynamics (QED). The strange particles do exist as strongly interacting particles but in the field of strong interactions no one was able to build a theoretical model such as the starting point of QED. The study of all pictures taken by the Blackett group at the Jungfraujoch Lab was my first priority. Even if the scientific interest of Professor Blackett had moved onto a totally different matter, like the magnetic field of our planet and its structure, the strange particles and the reason for their existence were radicated in his scientific youth. Professor Blackett often invited me for lunch at his table at Manchester as well as at Imperial College to discuss scientific matters as well as other problems concerning the value of scientific research in everyday life. When I was in his office I realized that he could speak with top leaders at the governmental decision-level. I remember his phone calls with the Prime Minister and with the Chancellor of the Exchequer. His position in favour of the establishment of CERN was attracting my attention. I knew practically nothing of what was going on for this new laboratory in Europe, but later I realized the relevance of what I was the testimony of. There is no doubt: the consequence of what I was listening – having the privilege of being the pupil of Professor Blackett – is best expressed by C.S. Wu [20] (already quoted in Chapter IV): *"Without the engagement of P.M.S. Blackett and I.I. Rabi, CERN would not have existed."*

Photo 24: Chien Shiung Wu at Erice (1994) during a lecture on the discovery of parity and charge conjugation invariance violation in weak interactions.

② Not many people know the origin of the Time-reversal invariance operator. How it did happen that all physics phenomena theoretically described – since the discovery of Science – during many centuries – with the same time arrow of our life (from past to future) **suddenly** started to be formulated on the basis that all events occurring with time going from past to future had to be exactly equivalent as if the time was going from future to past?

Wigner and Johnny von Neumann were schoolmates and great friends. Wigner was convinced that Johnny was a genius and von Neumann was convinced that Eugene was a genius. They often used to discuss physics and mathematical problems losing their sense of time going on. During one such discussion, Wigner suddenly said: "*My God, it is already midday. I was invited for lunch. Please excuse me Johnny, I must go immediately.*" Wigner was an extremely polite fellow, very kind and always exactly on time. Rabi used to say: "*Kindness is the Wigner's weapon.*" As Wigner was running off, von Neumann said: "*Eugene, you are so smart; why don't you stop the running of time?*". And, immediately added: "*Sorry, since it is already too late, you should be able to let time go backwards.*" That started Wigner thinking about elementary particles and Fundamental Laws of Nature. An elementary particle has no brain, no watches: it cannot distinguish between past and future. All fundamental interactions among elementary particles must be invariant if time goes one way or in the opposite direction. Three months after the famous von Neumann invitation to his friend Wigner to use his intellectual power and imagination in order to have time going backward instead of forward, Wigner discovered the famous theorem which establishes the existence of the Time-reversal operator. All Laws of Physics should be, as we now say, "T Invariant", i.e. physics reality should remain the same if we invert the time flow. I have performed the first experiment in order to check if, in the electromagnetic interactions, "T Invariance" was valid. Wigner was very happy to know that electromagnetic forces obey his (Wigner's) Invariance Principle. Professor Wigner is one of the greatest scientists of all times. His intellectual power was admired by von Neumann who once told me something like the following: "*I could tell you all details of what I have done on 25 June 1933.*" He was convinced that his intellectual power and imagination (see "*The limits of Human Imagination*", Appendix 23, and "*Beyond the limits of Human Imagination*", Appendix 24) were not as great as he would have liked because he had too much memory. For him Eugene Wigner had a stronger intellectual power and much more imagination. Why? Because: "*Fortunately for him, his memory is not as powerful as mine.*" Wigner was convinced that von Neumann's intellectual power and imagination were stronger than his own. For example von Neumann could have discovered three years before what Gödel did in 1931. The most famous mathematician in the world, Hilbert, was promoting – the world over – a program for resolving the crisis of mathematics by solving the "**decision**" problem.

A formal method of deciding the truth or falsehood of every mathematical statement was needed. Hilbert declared that to resolve the crisis of mathematics it was necessary to find a set of **axioms** that were proven to be both **consistent** and **complete**.

Consistent means to never have a theorem and its negation.

Complete means that every statement can be proved to be right or wrong.

Mathematics could rest on a firm logical foundation if every meaningful mathematical statement could be proved true or false.

From 1925 to 1928, von Neumann was trying to rescue mathematics, as we will see in Appendix 14 "*Wigner, von Neumann and Gödel*". During these three years von Neumann created a simple and beautiful new set of axioms [110], which were later shown by Kurt Gödel to be exactly what was needed for understanding the true nature of mathematics. "*Johnny could have discovered what Kurt Gödel did*", Eugene Wigner once told me.

In 1931 Kurt Gödel in Vienna proved two theorems that devastated the Hilbert program. **Gödel proved** that no system of axioms for mathematics could be **complete** and that no system of axioms could prove itself to be **consistent**.

In another occasion Wigner reminded the students of the Subnuclear Physics School in Erice that it was Johnny von Neumann who succeeded in making Quantum Mechanics mathematically respectable. We should not forget that this was and still is the best proof on the mathematical validity of **Quantum Mechanics**.

Von Neumann was very impressed by the Wigner achievements. Wigner created the "Time-reversal operator" because he was able to show that this operation (i.e. to reverse the arrow of **Time**) does not produce any contradiction in the Logic of Nature. In 1964 a special effect was discovered which violates "T Invariance", in the decay of a particle now called K_L^0-meson (long-lived K^0-meson). This effect has its origin in the weak forces not in the electromagnetic forces which are indeed T Invariant, like the strong subnuclear forces and the gravitational forces. I must thank Professor Blackett for giving his research group on cosmic rays to CERN, where high energy machines were being constructed. This allowed me to use cosmic rays to discover the pair production of heavy strange mesons thus producing the so much wanted experimental evidence that Fermi wanted in order to believe in the existence of the "strangeness charge". From cosmic rays I could move to machine physics thanks to the first CERN accelerator SC (see *"Europe had only Cosmic Rays"*, Appendix 3). It is using SC that I did my first experiment to prove the validity of the Time-reversal invariance in QED, allowing me to know Wigner. It is thanks to Wigner that I had the privilege to know von Neumann and Kurt Gödel.

③ Probably very few people know that the greatest of all mathematical minds, Kurt Gödel, was a physicist of the famous "Vienna Circle" and that his discovery came one year after the Heisenberg Uncertainty Principle that attracted his attention. The Vienna Circle wanted to give top priority to **Physics** in all problems to be understood. This means that the most important discovery in rigorous mathematical logic, i.e. the **Mathematical Impossibility to Decide** if a theorem can be demonstrated to be true or false, is due to a young physicist who was very much impressed by the **Uncertainty Principle** discovered by Heisenberg in 1930. I learned how Heisenberg discovered his principle, when at CERN in 1965 my group discovered the existence of Antimatter [101].

This discovery allowed me to personally meet Heisenberg and to know from him that the first conference he went to in his life was when a cloud-chamber picture of the electron in cosmic rays was presented by Wilson, the inventor of the cloud-chamber (Chapter VI, **Photo 9**). At the same conference Niels Bohr discussed his theory about the non classical property of the "electron" in an atom. Einstein was at the Conference and during the coffee-break he saw a very young fellow, Heisenberg.

Photo 25: The young Heisenberg when he was interested in knowing the Physics frontiers of the time.
(Photo kindly given by his family).

"*Have you seen* – Einstein asked to Heisenberg – *the photo shown by Wilson of the electron? A perfect classical trajectory obeying the Euclidean geometry. How can it be that a particle following a "classical" trajectory behaves in atoms as imagined by Bohr? I asked Bohr and he told me that this is an irrelevant detail.*" Heisenberg was at the Conference replacing his professor,

Sommerfeld, who could not attend, being very busy with his work on the 3rd motion of the Earth which apparently was not following the theoretical expectation of the gravitational forces. Sommerfeld was teaching the young Heisenberg: when people say "*this is a detail*" the reason is that they have not understood the problem. From a "detail" – was Sommerfeld repeating to Heisenberg – totally unexpected phenomena can come out.

It is the Niels Bohr's "detail" which sparked in Heisenberg the interest to understand what the trajectory of the electron really is. The "classical" trajectory of the electron is, in terms of physics quantities, made of "pieces" of linear momentum (Δp) and pieces of space (Δx) where the electron moves. The product of these two quantities is an "action". The same trajectory is also made with pieces of "energy" (ΔE) which the electron has and pieces of "time" (Δt) during which this energy exists. The product of "energy" and "time" is again an "action". Planck in 1900 discovered that the smallest amount of "action" cannot be as small as zero. The minimum action is the so-called "quantum of action" now known as "Planck's constant". A few months after the Conference, Heisenberg discovered what is now known as the "**Uncertainty Principle**". In fact if you increase your knowledge on one quantity which makes up the action – in the trajectory – you lose the knowledge of the other quantity, which makes up the same action in the trajectory. The **Uncertainty Principle** starts with a remark by Einstein to the young Heisenberg. And this originated the **Mathematical Impossibility for Decision-Making Actions** discovered by Gödel thanks to the "Vienna Circle". It is probably interesting to recall what Gödel did.

For thousands of years all human intellects had thought that only two possibilities can exist for a statement (a theorem is a statement): either it is true, or it is false. A third possibility cannot exist. This is the famous principle of the **excluded third**. Gödel discovered instead that in the heart of axiomatic logic there is the **third possibility**. Namely, that a theorem will always exist for which it will not be possible to decide whether it is true or false. Note: Gödel did not discover that a theorem can be both true and false.

Photo 26: Master Arnold Sommerfeld (1868-1951) with his "pupil" Werner Heisenberg (1901–1976), two giants of Science. When Sommerfeld went retired, the Direction of the Institute for Theoretical Physics in Leipzig was not entrusted to Heisenberg. Reason: the Nazis imposed their veto, mindful of what Heisenberg had dared to do when all Jewish professors were expelled from the University. In a memorable speech Heisenberg tried to convince his colleagues of the Faculty to denounce the expulsion of Jewish professors calling it immoral, as Planck had already done with Hitler. The successor of Sommerfeld chosen instead of Heisenberg was Professor Wilhelm Müller called by Sommerfeld a "*total scientific nullity*".

(*Photo courtesy of the family Heisenberg*).

Gödel discovered, one year after the **Uncertainty Principle of Heisenberg**, that the **principle of the excluded third** is wrong, after thousands of years of absolute certainty as to its validity. The origin of these two great conquests in **Physics** and in **Mathematics** is in a telegraphic comment, by Einstein, to the young Heisenberg.

④ We have said in Chapter III-3 that the so-called Lamb-shift has been experimentally discovered and theoretically computed by Willis Lamb. We have also pointed out that Weisskopf was the first physicist who computed a much more difficult theoretical effect, the so-called "vacuum polarization" in hydrogen, having taken very seriously the discovery by Blackett and Occhialini in 1932 of the (e^+e^-) production in cosmic rays, which thanks to Dirac gave rise to the effect computed by Weisskopf.

Being the Lamb-shift the simplest "virtual phenomenon" how did it happen that Weisskopf was not the first to perform the theoretical calculation of the Lamb-shift, having been able to compute the much more difficult "vacuum polarization" effect on the hydrogen atom. Here is the sequence of events how they really did happen.

When Lamb announced his discovery, Weisskopf with his student J. Bruce French were the first to correctly calculate the Lamb-shift. Julian Schwinger and Richard Feynman were also engaged in the same calculation but, both of them had made the same mistake, thus getting the same answer (wrong). Unfortunately, both Schwinger and Feynman were in contact with Weisskopf who could not believe that these two members of the younger generation of physicists engaged in computing this unpredicted new effect could both be wrong. Thus Weisskopf decided to postpone the publication of his result. Schwinger and Feynman meanwhile found their mistake but Lamb and his student Norman Krool published their result while Weisskopf was weighting for the cross check. There is an interesting detail to be pointed out. The Lamb-shift can be computed within 5% without using relativistic calculation (as done by Hans Bethe). But there is another effect: the magnetic behaviour of the electron. This corresponds to the effect caused by the emission and reabsorption of virtual photons on the magnetic moment of an electron moving in a magnetic field. The magnetic moment of the electron and its g-value is the result of the relativistic description of the electron by the Dirac equation. The fact that the gyromagnetic ratio of the electron is predicted by the Dirac equation to be $g = 2$ cannot be accounted for by any non-relativistic description. The "anomalous" magnetic moment of the electron, i.e. its g-value being different from 2 needed a relativistic description of the virtual electromagnetic processes. Furthermore, while the Lamb-shift affects only the hydrogen atom, the deviation from QED of an intrinsic property, such as the "magnetic" moment of an elementary particle (the electron), is expected to affect other particles as well. In fact the anomalous magnetic moment of the muon was considered the crucial check in order to verify the intrinsic property of this particle (as discussed in Chapter III-3.1).

⑤ When Einstein discovered in 1905 his most famous equation

$$E = mc^2 \qquad (1)$$

he could not sleep during three months (Peter G. Bergmann testimony). During many decades before 1905 the letter "m" was used to mean mass and matter. Thinking about what could the letter "m" means, Einstein realized that mass and matter are two different physical quantities. In fact J.J. Thomson in 1897 discovered the "electron". Matter, in addition to its weight carries an enormous number of electric charges. Having realized that "m" in his equation corresponds to "mass" not to "matter" Einstein could sleep again. Some historians are convinced that, if the electron had not been discovered before 1905, Einstein would not have published his equation (1). The historians recall the introduction of the famous Λ (the vacuum energy), introduced "ad hoc" in his equation for the Universe in order to avoid the expansion of the Universe, since he was convinced that the Universe was static. The expansion was discovered in 1929 by Hubble; Einstein did not like to propose theoretical structures in contradiction with what was believed to be correct in everyday life.

Back to real life. A glass of water is made with molecules which consist of two hydrogen atoms electromagnetically bound to one atom of oxygen. The molecule H_2O needs 10 electrons, 10 protons and 10 neutrons. In the mass of 100 grams of water there are billions of billions of electrically charged particles, electrons and protons. This means that if "m" in equation (1) would mean "matter" all electric charges should disappear in order to become "energy". Since the electric charges have to obey a law which forbids their disappearance the letter "m" in equation (1) cannot represent "matter", but only the "mass" of the water. This is why in reaction (1) "m" stands for "mass" nor for "matter". The Einstein equation (1) tell us that if we have 100 grams of matter, nothing is going to happen thanks to the "stability" of matter linked to the fact that a proton can never become an electron. The proton has positive charge, the electron negative.

In 1929 Dirac with his equation discovered that not only negative electrons exist but also positive electrons (e^+). Why the proton does not become a positive electron? As repeatedly emphasized, the existence of e^+ was established by Anderson, Blackett and Occhialini in 1932. A proton with positive electric charge can become a positive electron, since the mass of the proton is 2000 times larger than the mass of the antielectron (e^+). And here comes the problem of the stability of the matter, any type of matter including water.

In 1938 Ernst Carl Gerlach Stueckelberg noted that the number of protons and neutrons (the heaviest particles in all types of matters familiar to us) in the Universe can never change, otherwise matter would be unstable. This is how the conservation of the "baryon number" was introduced in Physics. Each nucleon (proton, neutron) has the intrinsic property called "baryon number" and this number must be conserved. This conservation law became of great interest in the seventies with the advent of the Grand Unified Theories (GUT), where not only the baryon number (B) but also leptonic number (L) conservation law had to be introduced. For details on these topics see [103] and Harry J. Lipkin "*SU(5) without SU(5): Why B-L is conserved and baryon number not in unified models of quarks and leptons*" in [Proceedings of the 1980–Erice Subnuclear Physics School *"The High Energy Limit"*, Vol. 18, page 281, A. Zichichi (ed), *Plenum Press*, New York-London (1983)].

⑥ How did it happen that I was able to realize projects and Institutions which, when proposed, appeared dreams and whose positive impact is still nowadays extremely relevant? The first example concerns what is now the most powerful hadronic collider in the world: the LHC at CERN with a 27 km circumference and a cross-section for the tunnel to allow not a single collider but two colliders to be installed. How it was avoided that a circumference of only 13 km and a small cross section for only one single collider is in Appendix 25: *"The Roots of LEP and LHC"*. The other examples are reported in the "scientific credibility" of the new Manhattan Project (*"A New Manhattan Project for the life and dignity of all people in the world"*, Appendix 21).

Conclusive Remarks.

Things do not become real as described by those who try to reconstruct the origins of all actions using Committees and Committees. Let me recall Fermi's statement: "*Committees have never discovered nor invented **anything**.*"

When I became a pupil of Professor Blackett I did realize that the personal direct interactions are needed at the decision-making level. Reports and papers needed to explain and reconstruct the basis for actions come after, often invented by historians, who try reconstructing what later becomes the history.

The intellectual power of convincing a fellow, at the decision-making level, is the condition, necessary and sufficient, for something you want to do to become real. And the origin of what becomes real can only be found studying the original scientific papers.

A final Note which refers to the origin of this volume.

On many occasions I have been asked, during two decades (1955–1975), to recall the first steps of CERN in physics and technology by all the CERN DGs from Cornelius J. Bakker (1955–1960) to John B. Adams (1960–1961), Victor F. Weisskopf (1961–1965), Bernard P. Gregory (1966–1970) and Willibald K. Jentschke (1971–1975). I would like to recall two other physicists, Professors Peter Preiswerk and Wolfgang Paul, who encouraged me to write the memories of my youth of being the pupil of Blackett.

The first time I knew about Paul is when I was studying the problem of constructing high precision polinomial magnetic field and Weisskopf came with Wolfgang Pauli, to visit my set-up. Pauli was the Pope of Theoretical Physics and Weisskopf had been his assistant. Pauli liked to say that his presence in a lab was very dangerous: the "Pauli effect" would have caused serious troubles in the technological structures. Nevertheless he wanted to see how high precision magnetic fields could be constructed 10^2 times faster and 10^2 times less expensive. He liked to be witty and cheerful; talking to me he said: "*Do not worry for the "Pauli effect" since I have also a real part, Wolfgang Paul.*" Many years later when the real part of Pauli came at CERN I told him that the great Wolfgang Pauli (**Photo 27**) had already introduced me to his knowledge. He told me that he knew my great master Blackett whose actions of support for the establishment of CERN had to be highlighted by his pupil.

Photo 27: Wolfgang Pauli and Chien Shiung Wu (Archive-Pauli, CERN).

Wolfgang Paul has been a strong supporter with John Bell, Werner Heisenberg and Victor Weisskopf, of my nuclear antimatter experiment [101]. He came at CERN in 1964 as Scientific Director of the Nuclear Physics Division, at the right time when CP invariance was experimentally discovered to be broken [28] as discussed in Chapter III-1. In the **Photo 28** Professor Paul delivering a lecture.

Photo 28: Wolfgang Paul at the Subnuclear Physics School (1973 Erice).

Preiswerk in 1950 was in Florence when Rabi proposed the establishment of a European high energy physics laboratory. He suggested two possible sites for the lab, one near Geneva (recommended to be the best) and one near Basle, in Alsace. He played an important role in bringing this proposal to become real. He knew the role played by Blackett to let UK be a strong supporter of CERN. Preiswerk was at the University of Berlin when Max Planck and Erwin Schrödinger, were giving lectures. He collaborated with Frédéric Joliot and Irène Joliot-Curie, shortly after they had discovered artificial radioactivity. He was a great expert on slow neutron physics – a field opened by Fermi in Rome – reaching two relevant results: 1) neutrons in hydrogenous materials reach thermal equilibrium; 2) neutron resonance absorption is function of the neutrons velocity. If this was known to the Fermi group in Rome the famous "irreproducible" result would have become "reproducible". Preiswerk was a very interesting source of information on the four years preceding the formal existence of CERN – **29 September 1954** – and on many famous physicists.

All I can do now is to express, to all who encouraged me, my gratitude for their interest on the roots of CERN and apologize for having done it so late. It is on April 2014, during my visit at the Imperial College, that I realized, as stated on the first page of this book, why we should – at least once in our life – stop doing all urgent matters we are bound with our work and give time to recall how things really happened in the activity we have been engaged.

How we physicists start to put down the exact formulation of our new ideas has been reported in the present chapter on the basis of what I could remember. These memories corroborate what John Wheeler elaborated after many long successful decades of activities in Physics: when a new idea comes in, we physicists should not start writing formulae but translate the new idea in terms of very simple effects to be imagined. Formulae must come later.

In 1979 John Wheeler gave a series of lectures in Italy at the Fermi School in Varenna [J.A. Wheeler, *"Frontiers of Time"*, SIF, Bologna, Italy, North-

Holland, Amsterdam (1979)]. On page 11 he writes: *"It is preposterous to think of the laws of physics as installed by a Swiss watchmaker to endure from everlasting to everlasting when we know that the universe began with a big bang. The laws must have come into being."* In *"The Ten Challenges of our Physics"* (Appendix 19) there is nothing about the mechanism which describes how a Fundamental Force of Nature, a Law, comes into being: the problem raised by Wheeler. This mechanism should indeed be studied; it was a problem in the discussions with Blackett and his friend Bertrand Russell. After many decades it has been abandoned, since no-one has been able to contribute towards a description of how a Fundamental Force is generated. On page 11 Wheeler continues: *"Therefore they could not have been always a hundred percent accurate. That means that they are derivative, not primary."* And on page 44: *"Of all strange features of the universe, none are stranger than these: time is transcended, laws are mutable, and observer–participancy matters."*

I started my physics activity studying if a new charge called "strange" was really existent. The "strange charge" opened the horizon of the Subnuclear Universe. No-one could have imagined the strangest features of the Universe where we live and which we are part of. These "strange" features must have their roots on "THE WHOLE OF OUR KNOWLEDGE" (Figure 31 in Chapters VI and X) and on the fact that – as repeatedly emphasized – Physics is the mother and the queen of all Sciences.

I would like to close this book renewing my deep sense of gratitude to my great teacher Pat Blackett, whose philosophical discussions with his friend Bertrand Russell are still very timely.

Few words to the reader

The book covers a century of fundamental scientific discoveries in Physics, told in the simplest possible way.

Professor Blackett was telling us – young fellows – that it is much more difficult to explain the scientific achievements using simple words than writing formulae. I have attempted to explain topics at the frontier of our most advanced scientific knowledge as they came mixed with problems and facts of everyday life.

John Wheeler was convinced that in Scientific Culture one of the biggest problems is the number of times a book written in order to let people know about discoveries and achievements, needs to be read; certainly no less than three times. The other big problem was focused by a great mathematician, philosopher and famous educator, John Kemeny. He was convinced that for Science to become a strong component of Modern Culture the authors of the original papers should contribute directly to explain the meaning of their achievements. The basic original papers are in fact discussed by very many people but read by very few (see Appendix 13 *"The Vienna Circle and Gödel's discovery"*.

My conclusion is that the first reading should include all Appendices and should go on without any worry about what could seem to be difficult. This will allow a general view before starting the second reading of my attempt to cover, in the simplest possible way, a century of discoveries in trying to understand the Logic of Nature.

The third reading should follow the sequence that the reader thinks to be the best for him to understand as much as possible my attempt of letting the queen of all Sciences to become part of our Culture, called Modern, but in fact pre-Aristotelian, since neither the Rigorous Theoretical Logic (Mathematics) nor the Rigorous Experimental Logic (Physics) are basic components of the Culture of our time.

APPENDICES

INDEX

APPENDIX 1:	*(quoted on pages 9 and 78)* Mach died convinced that Atomic Physics was not Science	189
APPENDIX 2:	*(quoted on pages 10, 12 and 104)* Fundamental Forces	190
APPENDIX 3:	*(quoted on pages 13 and 175)* Europe had only Cosmic Rays	191
APPENDIX 4:	*(quoted on pages 17 and 104)* Fermions and Bosons	193
APPENDIX 5:	*(quoted on page 17)* Supersymmetry and Superworld	197
APPENDIX 6:	*(quoted on pages 27 and 39)* Asymptotic Freedom and Confinement	200
APPENDIX 7:	*(quoted on page 29)* Weak Interactions	201
APPENDIX 8:	*(quoted on page 33)* The Effective Energy	203
APPENDIX 9:	*(quoted on page 40)* The End of a Myth	205
APPENDIX 10:	*(quoted on pages 42 and 76)* The Yukawa's meson: the first example of the nuclear glue	206
APPENDIX 11:	*(quoted on page 47)* Virtual Physics and the Annihilation Processes	208
APPENDIX 12:	*(quoted on page 78)* Why we cannot see an atom	211
APPENDIX 13:	*(quoted on pages 78, 81, 82, 84 and 186)* The Vienna Circle and Gödel's discovery	212

APPENDIX 14:	*(quoted on pages 81, 83 and 174)* Wigner, von Neumann and Gödel	218
APPENDIX 15:	*(quoted on page 83)* Immanent and Transcendent	220
APPENDIX 16:	*(quoted on page 83)* Creativity in mathematical logic: Infinity	221
APPENDIX 17:	*(quoted on page 96)* Why electrons are so important	222
APPENDIX 18:	*(quoted on page 102)* The problem of going from inert matter to living matter	223
APPENDIX 19:	*(quoted on pages 112 and 185)* The Ten Challenges of our Physics	224
APPENDIX 20:	*(quoted on page 114)* Ettore Majorana: Genius and Mystery, what Robert Oppenheimer told me	225
APPENDIX 21:	*(quoted on pages 123 and 181)* A New Manhattan Project for the life and dignity of all people in the world	227
APPENDIX 22:	*(quoted on page 150)* The Useless Particles and the Origin of the Universe	230
APPENDIX 23:	*(quoted on page 174)* The limits of Human Imagination (Galileo Galilei)	232
APPENDIX 24:	*(quoted on page 174)* Beyond the limits of Human Imagination	233
APPENDIX 25:	*(quoted on page 181)* The Roots of LEP and LHC	234
APPENDIX 26:	*(quoted on Appendix 1)* Paul Dirac, Antimatter and 'Virtual Phenomena'	235
APPENDIX 27:	*(quoted on Appendices 11 and 21)* The Virtual Annihilation Phenomena open the horizon for the Grand Unification	236
APPENDIX 28:	*(quoted on Appendix 13)* Hilbert and the German Emperor	237
APPENDIX 29:	*(quoted on Appendix 13)* The basic achievements of Rigorous Theoretical Logic	238

APPENDIX 1: *(quoted on pages 9 and 78)* **Mach died convinced that Atomic Physics was not Science**

Ernst Mach (1838–1916) died convinced that his friends and colleagues had gone wrong. In fact, it was the period in which the atomic theory of the structure of matter first began to be developed.

Mach could not reconcile himself to the idea that no one would ever be able to see or touch these newly fashionable entities called atoms which are not visible to our eyes (see "*Why we cannot see an atom*", Appendix 12).

Science should not be devoting itself to entities of this kind, he thought.

Mach was forgetting Galilei. He would have said to Mach: forget about whether or not atoms are visible or touchable. What counts is knowing whether it is possible to imagine, and hence to realise, experiments with them that are capable of giving rigorous and reproducible results.

Atoms were grounded on hypotheses whereby these new entities must have dimensions such that they could neither be seen nor touched. These properties would not have worried Galilei. He considered experimental proof capable of providing reproducible results to be essential. Galilei would have demanded reproducible experimental proof from those who supported atomic theory.

Mach died convinced that **atomic theory** should be abandoned, thereby forgetting Galilean teaching: rigour and experimental reproducibility.

Today we work with "virtual phenomena" (see "*Paul Dirac, Antimatter and 'Virtual Phenomena'*", Appendix 26), which produce reproducible effects on properties such as the anomalous magnetic moment of the muon (the first high precision experiment at CERN, discussed in Chapter III and Section III-3.1).

APPENDIX 2: *(quoted on pages 10, 12 and 104)* **Fundamental Forces**

In our Universe there are four Fundamental Forces: Gravitational, Electromagnetic, Weak Subnuclear and Strong Subnuclear.

1) The Gravitational Forces help us to understand why stones fall from high to low and why our Earth is bound with our Sun; why our Sun is bound to our Galaxy and why cosmic structures exist in the Universe.

2) The Electromagnetic Forces explain the existence of light, TV, radio, electronic apparatuses, internet and mobile phones plus all other technologies familiar in our everyday life.

3) The Weak Subnuclear Forces, now called Fermi Forces, tell us what the stars are made of and why our Sun can shine for ten billion years without going out or exploding (for a more elaborate description of the Weak Subnuclear Forces see "*Weak Interactions*", Appendix 7).

4) The Strong Subnuclear Forces explain how the internal part of an atom can exist and be made of protons and neutrons. These two particles have in their structure the Subnuclear Universe. This fascinating Universe is in each piece of matter including our body.

The Electromagnetic and the Weak are mixed at the Fermi energy. This is why these two forces are also said to be only one: the Electroweak. For this reason the total number of forces is often said to be only three.

APPENDIX 3: *(quoted on pages 13 and 175)* **Europe had only Cosmic Rays**

Before 1957 – the year when the Blackett group working with cosmic rays at CERN discovered the pair production of heavy mesons with positive and negative "strange" charge – there were no artificial sources of cosmic rays.

The first machine producing protons with the top energy of 600 million electron-Volts was SC, as reported in **Table 1**.

---

① **SC (CERN Synchrocyclotron) 1957 – 1990**
  - **top energy:** 600 MeV (proton on fixed target)

② **PS (Proton Synchrotron) 1959 – present:**
  - **top energy:** 28 GeV (proton on fixed target)

③ **ISR (Intersecting Storage Rings, the first proton-proton collider) 1971 – 1984**
  - **top energy:** 62 GeV (in proton-proton collisions)

④ **SPS (Super Proton Synchrotron) 1976 – present:**
  - **top energy:** 450 GeV (proton on fixed target)

⑤ **LEP (Large Electron Positron Collider) 1989 – 2000:**
  - **top energy:** 209 GeV (in electron–positron collisions)

⑥ **LHC (Large Hadron Collider) 2009 – present:**
  - **top energy**: 13 TeV (in proton-proton collisions)

Table 1

In everyday life, the energies we are dealing with are at the level of a fraction of electron-Volt.

From 1957 to now during these nearly six decades, Europe has constructed six big machines able to reach now (2015) the top energy of 13 TeV.

> Note
> - one million electron-Volts ≡ 1 MeV
> - one thousand MeV ≡ 1 GeV
> - one million MeV ≡ 1 TeV.

As reported in **Table 1** CERN accelerators started in 1957 at the total nominal energy of the incident proton equal to **600** million electron-Volts (600 MeV). Now CERN is colliding beams at nominal total energy of 13 billion electron-Volts (13 TeV) which is the highest in the world.

With the first two machines SC and PS the collisions studied were produced by the protons of the accelerators against protons at rest in the Laboratory (the technical terms was: protons against fixed target).

Victor Weisskopf decided to jump, from using one beam of protons against fixed targets (SC and PS), to the first collider where two beams of protons were colliding one against the other. The collider was named ISR (Intersecting Storage Rings). The jump in energy is due to the fact that if you bombard a fixed target with protons, the protons in the target are at rest. All the energy of the collision is in the incident proton. When two protons collide, as it is at ISR, and in the other colliders (LEP and LHC) the total amount of energy available for the study of what happens is the sum of the energies of the two beams: $2E_{beam}$.

If the collision is of a proton of 28 GeV against a fixed target the total energy is 7.5 GeV much smaller than 56 GeV, the energy obtained when two beams of 28 GeV collide.

APPENDICES

APPENDIX 4: *(quoted on pages 17 and 104)* **Fermions and Bosons**

The world familiar to us is made of elementary building blocks (six quarks and six leptons) plus four types of glues, the gravitational, the electromagnetic, the weak and the strong. The building blocks and the glues follow two drastically different statistical laws. And this means that, if you have many building blocks of the same type and many glues of the same nature, the way they stay together is very different indeed. In a Lecture Hall, if there are 1000 seats, you cannot expect to seat one million fellows. Each fellow needs his seat. This corresponds to Fermi-Dirac statistics.

There is another law, called Bose-Einstein, statistic. According to this law, if the fellows attending a Conference were one million, they could all be accommodated in the Lecture Hall where there are only one thousand seats.

Every object you can think of must obey either the Fermi-Dirac or the Bose-Einstein statistics. Does an object exist which obeys partly one and partly the other? The answer is no. Objects obeying the Fermi-Dirac statistical law are, for brevity, called fermions, while objects obeying the Bose-Einstein statistical law are called bosons. Either an object is a "fermion" or it is a "boson".

For us physicists the objects are neither the fellows attending a Conference in a Lecture Hall, nor complicated objects such as a stone or a flower. Our objects are "elementary". And the elementary objects obeying the Fermi-Dirac statistical law are the fundamental **building blocks** of our world: quarks and leptons. These **objects** are all **fermions**.

193

The other class of elementary objects obeying the "Bose-Einstein" statistical law is that of the fundamental **glues** quoted above. These objects are all **bosons**.

How do we distinguish a fermion from a boson? What is the basic property which allows physicists to be sure that an elementary object is either a fermion or a boson? This basic property is called spin. **Building blocks** and **glues** may be thought of as being very small pieces of matter. These very small pieces of matter must then be imagined as being in constant motion, like a top. The name given to this intrinsic property (the spin) of an elementary object – being it **building block** or **glue** – has its origin in the English name used to specify the motion of a top. The **spin** of a particle is as intrinsic as its **mass** and its **charge**.

The spin of a particle is measured in units of the Planck constant. The smallest amount of spin is (1/2) of the Planck constant. All quarks and leptons have spin equal to (1/2). All glues have spin equal to **one unit** of the Planck constant.

We see here the fundamental distinction between building blocks and glues.

The building blocks have semi-odd-integer values for the spin.

The glues have integer values for the spin.

It is not an irrelevant detail.

All objects carrying semi-odd-integer values of spin, 1/2, 3/2, 5/2, etc., are called fermions.

All objects carrying integer values of spin, ZERO, 1, 2, 3, ... etc., are called bosons.

Back to our previous example. Imagine a theatre with one thousand seats. If you have fermions as spectators you are forced to give one seat to each Mr. Fermion and therefore – as said before – you cannot dream of selling more than one thousand tickets.

But if you have bosons you can sell ten thousand million tickets. In a seat you can put not only one but as many Mr. Bosons as you like.

We physicists do not work with theatres, seats, Mr. Fermion and Mr. Boson.

Our Mr. Fermions are quarks and leptons. And the Fermi-Dirac statistical law says that in each seat (better known as a "position in the phase-space") we cannot have two identical fermions. With Mr. Bosons the Bose-Einstein statistical law says that we can have (in each "position in the phase-space") as many identical bosons as we like.

This introduction is needed because what is called "common-sense" has its roots in our daily life. For us the existence of bosons would be unthinkable. Unthinkable but real. By real I always refer to what Galilei taught us: real is whatever you can imagine, provided it gives rise to reproducible, rigorous experimental results.

Superfluidity and superconductivity are examples of the bosonic properties of real objects.

Why is it that in a tube of given cross-section, the amount of water which you can hope to have passing through cannot exceed a well-known quantity? What fixes this limit? Answer: the friction between the molecules on the surface of the tube and the molecules of the running water. The external part of each molecule is the electron cloud. Electrons are fermions. If you could produce a liquid which is made of bosons, then the amount of this liquid going through a tube would be as much as you like. This is superfluidity, discovered by Pyotr L. Kapitza many decades ago.

An analogous phenomenon takes place in superconductivity. Given an electrically conducting cable, the reason why the amount of electric current through this cable cannot exceed a given value has its roots in the properties of the "electron". The electric current traversing a wire is due to electrons. And we know that electrons are fermions. Why? Because their spin is ½. Imagine two electrons together. The system composed of two electrons has spin **1**, which is one-half plus one-half. If the "**electron pair**" (called also the Cooper-pair from the name of the physicist who understood superconductivity) can be considered as a **unique entity**, this entity, made with two fermions, behaves as a boson. In this case you can put in the same place as many "electron pairs" as you like. And this is how superconductivity works. We will see later that there are two types of superconductivity. One is the phenomenon understood by John Bardeen, Leon N. Cooper and John R. Schrieffer, and refers to what is called "cold" superconductivity, i.e. the superconducting property of a cable exists insofar as the cable is maintained at a very low temperature. More recently "warm" superconductivity has been discovered by two experimental physicists, K. Alex Müller and J. Georg Bednorz of the IBM Lab in Zurich. Here the wire can be warm. The explanation is always in terms of "electron pairs". In the case of "cold" superconductivity, the two electrons are electromagnetically bound at "large" distances. In the "warm" superconductivity the two electrons are electromagnetically bound at "small" distances. The "warm" superconductivity has no complete theoretical interpretation. The most elegant one is due to T.D. Lee who has suggested further experiments in order to find out the complete answer to this fascinating new effect.

One conclusion is in order. The real world has experimentally reproducible effects which prove that both statistical laws (the Fermi-Dirac and the Bose-Einstein) are operative in Nature.

Thus our common sense once again turns out to be very limited indeed, and our imagination too, as proved by superfluidity and superconductivity.

APPENDIX 5: *(quoted on page 17)* Supersymmetry and Superworld

Supersymmetry is a new property of Nature which puts on equal basis Fermions and Bosons.

Our world seems to indicate a clear-cut distinction between fermions and bosons.

Fermions are building blocks. Bosons are glues.

This clear-cut distinction could be another example of common sense. The question is in fact as follows: what is the fundamental reason for which **fermionic** property **must** belong to the building blocks, while **bosonic** property **must** belong to the glues.

The answer is that this fundamental reason is nonexistent. The more you think about this problem, the more you become convinced that Nature should be totally symmetric. It should not attribute any privilege either to the fermions or to the bosons.

In other words the building blocks can exist with the property of being fermions and also with the property of being bosons. Consequently, the quarks and leptons of our world happen to have spin ½. This is only half the truth. The other half is the following: quarks and leptons with spin zero must also exist. To these – so far hypothetical – new particles, the names of superquarks and superleptons have been given. If it is true – and we believe it is – that the electron is there, the superelectron should also be there. Where, we will see later.

The same symmetry applies to the glues in our world, which have the properties of being bosons. If Nature obeys the boson-fermion symmetry (called Supersymmetry), the existence of glues which have the property of being fermions is necessary, compulsory even. One example of glue in our world is the object called a photon: the light we see with our eyes is made of photons. These photons have spin **1**. If Supersymmetry is a Fundamental Invariance Law of Nature, we must also have photons with spin ½ . The name given to the light with fermionic properties is superlight or photino.

In our world we have electrons and light. If Nature is supersymmetric, sooner or later the superelectron and the superlight should be discovered. And this is not all. Supersymmetry extends its validity to all leptons and all quarks, as well as to all glues. Superleptons, superquarks and superglues make up the Superworld.

What about Space-Time?

Are the properties of Space-Time fermionic or bosonic? The answer is bosonic. The reason why we can go back and forth in space is due to the bosonic nature of the space-dimensions. In Time we cannot go backward but only forward. The reason is unknown. It could be because we are complicated structures, made of billions and billions of elementary objects. Eugene Wigner has demonstrated that if an elementary particle interacts with other elementary particles, no matter what time-arrow you choose, the result must be the same. I did an experiment, many decades ago, to check the validity of Wigner's theorem on Time-reversal invariance (Chapter XII, point n. 2) in the electromagnetic interactions. The experimental results confirm with high accuracy (the highest so far achieved) that Eugene Wigner's theorem is indeed valid in the realm of all electromagnetic interactions. In other words if we are unable to go backward and forward in time, this is not due to the intrinsic property of time. Time is in fact bosonic. This is why, in our experiment, reproducible results are obtained showing that Time-reversal invariance indeed holds when electromagnetic forces act between elementary particles.

In our world the nature of Space-Time is bosonic. What about a fermionic Space-Time? Can such an entity exist? The answer is yes. It is called Superspace. In Superspace also you can imagine going from one point and back to the same point. However, when you get back, you are not exactly as you were when you started to move.

The reasons why I believe in all these fascinating speculations – like the Superworld – are manifold. The first reason is that what we call the "Standard Model" is not enough to give the answer to the questions posed by the experimental observations already available, as discussed in Chapters VIII and X. The Superworld with 43 dimensions is discussed in Chapter VI.

APPENDIX 6: *(quoted on pages 27 and 39)* **Asymptotic Freedom and Confinement**

"Asymptotic freedom" corresponds to the fact that when two quarks collide at very high energy their interaction decreases and they cannot attract each other, like they do at low energy.

In fact "confinement" corresponds to the property that quarks at low energy remain linked together, i.e. "confined".

Experiments prove that asymptotic freedom and confinement do exist. But, while asymptotic freedom is theoretically predicted by Quantum ChromoDynamics (QCD), confinement is at present not possible to be deduced from QCD.

APPENDIX 7: *(quoted on page 29)* Weak Interactions

Weak Interactions are not philosophical entities. The weak interactions are generated by the weak forces. The Sun burns so regularly for billions and billions of years thanks to the existence of these fundamental forces of Nature.

Our Sun, like any star we can admire in the sky, is a **mixture** of protons and electrons in exactly equal numbers. The electric charge of the proton is equal but opposite in sign to the electric charge of the electron. This is why the total electric charge of the Sun is zero. This enormous **mixture** of protons and electrons would remain inactive for billions and billions of years, if it was not for the existence of the fundamental force of Nature quoted above: the weak force. Thanks to this force the **mixture** of protons and electrons slowly develops another component of the mixture. This other component is made with two other types of particles, called neutrons and neutrinos. Imagine a proton plus an electron. In the Sun, thanks to the gravitational pull, these two particles acquire energy. Sometimes they collide.

Thanks to the weak force, the collision between a proton and an electron produces a neutron and a neutrino. The neutrino escapes immediately and reaches us. Every second, day and night, there are sixty billion neutrinos impinging on a square centimetre surface of my finger. They come from the Sun. As a neutrino has neither electric nor nuclear charge it can traverse matter, the entire Earth and even a great cement wall billions of kilometres deep, without interacting. At the Gran Sasso Laboratory we have very sensitive instruments able to observe these neutrinos. The Sun and all the stars continually emit neutrinos.

What about the neutron? This is essential for our life. The neutron, once produced inside the Sun, can interact with a proton and, thanks to the nuclear forces it forms a deuteron. Two deuterons together, thanks to the nuclear force which generates nuclear attraction, form a Helium nucleus. The mass of the two deuterons is larger than the mass of the Helium nucleus. This excess of mass transforms into an enormous amount of energy thanks to the famous law discovered by Einstein: $E = mc^2$. This transformation of mass (m) into energy (E) follows the same law which produces the so called H-bombs. The difference is in the fact that the Sun works like a candle, thanks to the weak force, while in the H-bomb, the transformation of mass into energy takes place at once since there is too much deuterium to become Helium. This enormous amount of energy is produced in the Sun's core and it takes one million years to reach the Sun's surface. The light which we see coming from the Sun is the result of the energy produced in the "Nuclear Fusion" of two protons plus two neutrons. If neutrons could not be produced, the Sun could not exist as a nuclear candle. The neutrons are the fuel of the Sun and the weak force is the safety valve which guarantees that the correct quantity of neutrons is produced every second in the (proton-electron) mixture of the Sun.

APPENDIX 8: *(quoted on page 33)* **The Effective Energy**

Imagine many fellows entering a shop, all having in their wallet different amounts of Euros. Suppose that in the shop you can buy only three kinds of objects: fruits, meat and bread. If a fellow buys one kilo of fruit he spends three times less than another fellow who buys three kilos of fruit. If a fellow buys all sorts of things he will spend all the money in his wallet. The money in the wallet we call the "**nominal money**". The money spent buying things we call the "**effective money**". In Physics we have strong, electromagnetic and weak forces: as if we can buy three kinds of things.

When a proton interacts with another particle, since the proton has all fundamental charges (strong, electromagnetic and weak) the interaction depends on the charges of the other particles. If the other particle has only the weak charge, the interaction can only be weak. Suppose that in the interaction of the two particles the proton, before the interaction, has a nominal energy of 100 GeV (100 billion of electron-Volts). This is like the fellow who has in his wallet 100 Euros. If we measure, after the interaction, the energy of the proton and we find that it is equal to 30 billion electron-Volts, this implies that the proton has spent 70 billion electron-Volts. This is the "**Effective Energy**" spent to produce whatever has been done in the interaction of the proton whose Nominal Energy was 100 billion electron-Volts. The proton Effective Energy could have been as high as the Nominal Energy. Not greater since there is a fundamental law of physics which establishes that the total energy must be conserved. It is the equivalent of saying that the fellow whose money in the wallet is 100 Euros cannot spend more than 100 Euros.

To find out what the effective amount of money is spent is easy. To find out the Effective Energy, of a proton or any other elementary particle, after the interaction is quite difficult.

This is the reason why all the interactions where elementary particles are involved have as the basic parameter the "Nominal Energy", not the Effective Energy.

When the interactions of different particles are compared using as a fundamental parameter the "Nominal Energy" the results are apparently dependent on the nature of the particles. As mentioned before these interactions could be strong, electromagnetic (EM) or weak.

No matter the nature of the interaction (strong, EM, weak) and the type of particle, if the **Effective Energy** is the same, all processes produce the same final states [38]. It is as if in our previous example, we would have discovered that, with the same amount of money spent to buy fruit, all fellows would have come out with the same kilos of fruit.

We discuss these problems in Chapter III-1.

APPENDIX 9: *(quoted on page 40)* The End of a Myth

THE END OF A MYTH: HIGH-p_T PHYSICS

"So far, the main picture of hadronic physics has been based on a distinction between high–p_T and low–p_T phenomena.

In the framework of parton model, high–p_T processes were the only candidates to establish a link between

- *purely hadronic processes*
- *(e^+e^-) annihilations*
- *(DIS) processes.*

The advent of QCD has emphasized in a dramatic way the privileged role of high–p_T physics due to the fact that, thanks to asymptotic freedom, QCD calculations via perturbative methods can be attempted at high–p_T and results successfully compared with experimental data [1]. The conclusion was: we can forget about everything else and limit ourselves to high–p_T physics.

Being theoretically off limits, low–p_T phenomena, which represent the overwhelming majority of hadronic processes (more than 99% of physics is here), have been up to now neglected. By subtracting the leading proton effects in order to derive the effective energy available for particle production and by using the correct variables, the BCF collaboration has performed a systematic study of the final states produced in low–p_T (pp) interactions at the ISR and has compared the results with those obtained in the processes listed below:

Process	Data Sources
(e^+e^-)	SLAC, DORIS, PETRA
(DIS)	SPS/EMC
(pp) ⎫ Transverse physics	⎡ ISR (AFS)
($\bar{p}p$) ⎭	⎣ SPS Collider (UA1)
(e^+e^-)	PETRA/TASSO (leading subtraction)

The results of this study [2-18] show that, once a common basis for comparison is found by the use of the correct variables, remarkable analogies are observed in processes so far considered basically different like

- *low–p_T (pp) interactions*
- *(e^+e^-) annihilations*
- *(DIS) processes*
- *high–p_T (pp) and ($\bar{p}p$) interactions*

This is how universality features emerge, and this is the basis to proceed for a meaningful comparison, i.e.:

<u>first</u> *identify the correct variables to establish a common basis,*

<u>then</u> *proceed to a detailed comparison**.*"*

---

* The root of this new approach to the study of hadronic interactions goes back a long time to a proposal by the CERN-Bologna group: "Study of deep inelastic high momentum transfer hadronic collisions" PMI/com-69/35, 8 July 1969."

Reproduction of the conclusions of a review paper [111].

APPENDIX 10: *(quoted on pages 42 and 76)* **The Yukawa's meson: the first example of the nuclear glue**

Yukawa in his 1935 paper [112] proposed searching for a particle with a mass between the light electron and the heavy nucleon (proton or neutron). He deduced this intermediate value – the origin of the name "mesotron", later abbreviated to meson – from the range of the nuclear forces. The search for cosmic ray particles with mass between those of the electron and of the nucleon became a hot topic during the 1930s thanks to this work.

On 30 March 1937, Seth Neddermeyer and Carl Anderson reported the first experimental evidence [113], in cosmic radiation, for the existence of positively and negatively charged particles heavier and with more penetrating power than electrons, but much less massive than protons. Then, at the meeting of the American Physical Society on 29 April, J.C. Street and E.C. Stevenson presented the results of an experiment [114] that gave, for the first time, a mass value of 130 electron masses (m_e) with 25% uncertainty. Four months later, on 28 August, Y. Nishina, M. Takeuchi and T. Ichimiya submitted to *Physical Review* their experimental evidence for a positively charged particle with mass between 180 m_e and 260 m_e [115].

The following year, on 16 June, Neddermeyer and Anderson reported the observation of a positively charged particle with a mass of about 240 m_e [116], and on 31 January 1939, Nishina and colleagues presented the discovery of a negative particle with mass (170 ± 9) m_e [117]. In this paper, the authors improved the mass measurement of their previous particle (with positive charge) and concluded that the result obtained, $m = (180 \pm 20)$ m_e, was in good agreement with the value for the negative particle. Yukawa's meson theory of the strong nuclear forces thus appeared to have excellent experimental confirmation. His idea sparked an enormous interest in the properties of cosmic rays in this "intermediate" range; it was here that a gold mine was to be found.

In Italy, a group of young physicists – Marcello Conversi, Ettore Pancini and Oreste Piccioni – decided to study how the negative mesotrons were captured by nuclear matter. Using a strong magnetic field to separate clearly the negative from the positive cosmic rays, they discovered that the negative mesotrons were not strongly coupled to nuclear matter [43]. Enrico Fermi, Edward Teller and Victor Weisskopf [118] pointed out that the decay time of these negative particles in matter was twelve powers of ten longer than the time needed for Yukawa's particle to be captured by a nucleus via the nuclear forces. They introduced the symbol μ, for mesotron, to specify the nature of the negative cosmic ray particle being investigated.

In addition to Conversi, Pancini, Piccioni and Fermi, another Italian, Giuseppe Occhialini, was instrumental in understanding that Yukawa had opened a gold mine. This further step required the technology of photographic emulsion, in which Occhialini was the world expert. With Cesare Lattes, Hugh Muirhead and Cecil Powell, Occhialini discovered that the negative μ mesons were the decay products of another mesotron, the "primary" one – "p" being at the origin of the symbol π [17]. This is in fact the particle produced by the nuclear forces, as Yukawa had proposed. It was this discovery which finally proved the existence of the first example of the nuclear glue.

APPENDIX 11: *(quoted on page 47)* **Virtual Physics and the Annihilation Processes**

The most spectacular consequence of the Dirac equation is the existence of the "annihilation" process. In fact the Dirac equation predicts the existence of the antielectron (also called "positron"). This implies, on one side the production of (e^+e^-) pairs by a photon and, on the other side, the annihilation of the (e^+e^-) pair into a photon (see Chapter III-3). These processes are governed by electromagnetic forces. The photon is the first "gauge boson" discovered. If other fundamental forces exist, the production of particle plus antiparticle and the annihilation into the appropriate gauge boson should take place. The name "gauge force" and "gauge boson" is due to Herman Weyl. Other gauge bosons exist when weak forces and strong forces come into play. When a particle (of any type) collides with its antiparticle, the pair (particle plus antiparticle) annihilates releasing their rest-mass energy as high-energy photons, or other gauge bosons.

For example, in the case of a process described purely by Quantum ElectroDynamics, a gamma-ray photon can create an electron-positron pair, which can transform itself back into a photon. This process, called "vacuum polarization" (see Chapter III-3), was the first "virtual" effect to have been theoretically predicted by Dirac in 1929 and experimentally proved to exist by Blackett and Occhialini in 1932.

The first physicist who computed the vacuum polarization effect in the hydrogen atom was – as discussed in Chapter III-3 – Victor Weisskopf. He predicted that the $2P_{1/2}$ level in a hydrogen atom should be very slightly higher in energy than the $2S_{1/2}$ level, by some 27 MHZ (MegaHertz). The unit used to measure the energy level of the hydrogen atom is MHZ. In 1947 Willis Lamb and Robert Retherford discovered that the $2P_{1/2}$ level was in fact lower than the $2S_{1/2}$ level by some 1000 ± 100 MHZ.

This experimental discovery that was to become known as the "Lamb-shift" effect prompted all theorists, including Weisskopf, Hans Bethe, Julian Schwinger and Richard Feynman, to compute the much more simple "virtual" process in which an electron emits and then absorbs a photon (see Chapter III-3). This very elementary process is the simplest example of the "virtual" phenomena. But it is the annihilation process which opens the great horizon of virtual phenomena. No one had the idea that this type of reality could exist before 1932, when Blackett and Occhialini discovered the simultaneous production by a γ–ray of an electron (e^-) and of its antiparticle (e^+). This experimental discovery gave rise – as said before – to the existence of (e^+e^-) annihilation back to a γ–ray. It is the virtual annihilation which allows a link between all fundamental forces of Nature as reported in *"The Virtual Annihilation Phenomena open the horizon for the Grand Unification"*, Appendix 27.

Let us try to see how we could imagine this "virtual" phenomenon. Think of a black box which contains a variety of objects. We cannot see the objects because the black box covers them entirely. If we put the box on a balance to measure its weight, we could deduce something about the objects inside the box. It could, for instance, happen that the weight changes from time to time. Sometimes the box is heavy, sometimes it appears lighter than in the previous measurements. Repeating the measurements, we could conclude that the box contains a bird-cage. If the birds are all flying while we measure the weight, the result is light. If all the birds are at rest in the cage, the corresponding measurement will give a weight higher than the previous one.

The "virtual" world in fundamental physics is much more fascinating than the box containing birds in a cage.

The Dirac equation is the mathematical instrument which allowed us to discover the "virtual" phenomena of the "virtual" world. In fact, if it had not been for the discovery of the positron – and therefore of the existence of the production of electron-positron pairs, as discovered by Blackett and Occhialini in 1932 – no one would have imagined that such "virtual" effects could exist in nature.

These topics are discussed also in Chapters III-3 and XII.

APPENDIX 12: *(quoted on page 78)* **Why we cannot see an atom**

An atom is tiny in comparison to light "balls".

We can only see structures whose dimensions are larger than those of light "balls".

A violet ball of light is four hundred nanometres (a nanometre is one thousandth of a millionth of a metre).

An atom is many thousand times smaller than a nanometre.

That is why we will never be able to see an atom.

APPENDIX 13: *(quoted on pages 78, 81, 82, 84 and 186)* **The Vienna Circle and Gödel's discovery**

The principal objective of the *Vienna Circle* was to *unify* all domains of *knowledge* **through the vehicle of Physics**. This "reductionist" conception met some success in domains like *Chemistry* and *Biology*, and it was this same premise that inspired Otto Neurath (1882–1945) and Rudolph Carnap (1891–1970) to advance the audacious thesis that every proposition must be expressible in the *Language* of Physics. For the *Vienna Circle* Physics had to be the mother of all Sciences.

These ideas came into debate among the physicists, mathematicians and philosophers of the *Vienna Circle*, which helped Kurt Gödel to become the greatest logician of all times.

The *Vienna Circle* ended up reaching a conclusion in line with Galilean thinking, according to which the only meaningful assertions are those that can be confirmed by a specified method or algorithm whose final goal has to be the result of a reproducible experimental proof in a laboratory. The reproducible experimental results were being achieved by the *Cambridge Circle* (see Chapter VI).

The "verifiability principle", based on the result of a reproducible experiment, is the basis of what has been called "**logical positivism**", a label that would come to be identified with the *Vienna Circle* and the *Cambridge Circle*. Not to be mistaken as **extraneous, this detail** in fact concerns one of the **most fundamental debates** in the *philosophy of Science*: is there a difference between the *truthfulness* of an assertion and its *demonstrability*? In other words, is it possible to **demonstrate** every **true proposition**? Demonstrability requires the existence of precise rules that can lead us to a conclusion of truthfulness in the final analysis. But can we be certain that such rules exist for all possible *truths*? This topic attracted the attention of mathematical logic, while the *Cambridge Circle* was concentrated on the experimental proof as the basic point.

Until 1931, all mathematicians would have agreed that such rules must exist for every possible truth.

In 1931, however, Kurt Gödel demonstrated that it is not always possible to couple a truth with its demonstration. By doing this, Gödel destroyed what had until then been a universal convention – that there was no difference between truthfulness and its demonstration.

Gödel discovered that there is an **unbridgeable** *gap* between what is true (within a logical system) and what we can effectively be **demonstrated to be true using the logical tools of that same system**.

Gödel's famous Theorem demonstrates the existence of fundamental limitations in the concept of demonstration itself.

This result is the greatest *Logical* **conquest of all times**.

This conquest puts tangible limits on human cognitive capacities for the first time in the history of thought. This **contradicted** the **monumental work** *Principia Mathematica* **by Bertrand Arthur William Russell and Alfred North Whitehead**.

The basic ingredient of *Principia Mathematica* (three volumes) was to eliminate "language" by using sequences of symbols to express all of the assertions of Classical Mathematics. To demonstrate that $1 + 1 = 2$, for instance, the exclusive use of **logical symbols required hundreds of pages**.

This was the price they thought necessary for meeting the challenge posed in 1928 by David Hilbert, the premier representative of mathematical thinking at the time of the International Congress of Mathematicians in Bologna.

The fundamental question posed by Hilbert in his address in Bologna, the most ancient and prestigious University in the world, was to determine whether or not it was possible to demonstrate all mathematical truths. Hilbert himself was firmly convinced that this was an obtainable objective: a system, rigorously logical and formally impeccable, that would axiomatize all Mathematics must exist. The challenge to the mathematical community re-visited the list of the 23 problems that the great German mathematician had presented in Paris in 1900 to mark the dawning of a new century (see *"Hilbert and the German Emperor"*, Appendix 28).

Hilbert's challenge met the most authoritative response in the work (1910–1913) of **Russell and Whitehead**, who believed that **freeing Mathematical Logic from the use of Language was the way out**.

Hilbert was seeking a sort of "truth machine", in which one could insert a proposition, push a button and receive an answer: true or false. Meeting the Bologna challenge instead of producing the "truth machine" shattered the link between what is actually true and what is logically demonstrable, a relationship that had provided a foundation for all mathematical thinking for centuries.

To briefly recall the history of all these steps can be instructive. The nineteenth century had already revealed how conceptions of "points", "lines" and "geometric figures" that had been held valid for thousands of years did not, in fact, describe **all of geometric figures**.

In the early 1800s, two scholars of antique geometric figures, **János Bolyai** and **Nikolaj Lobačskij**, independently discovered the existence of "triangles" for which the sum of the angles can differ from the famous 180 degrees of Euclidean geometry. In "hyperbolic" geometry, for instance, the sum of the angles is less than 180 degrees, whereas in "elliptical" geometry it is greater than 180 degrees. In the universe of "points", "lines" and "geometrical figures", these non-Euclidean truths are just as "true" as those of Euclidean geometry. In Euclidean geometry, given a straight line and one single point external to it, then there is one and only one straight line that is parallel to the given line. In the non-Euclidean geometries, alternatively, there can be either an infinite number of parallel lines, or no lines at all.

For the greatest exponent of mathematical thought of the times, David Hilbert, the **non-Euclidean geometries** had shifted the issue from **truth** and **demonstrability** (the sum of the angles that differed from 180 degrees in non-Euclidean geometries was rigorously demonstrable, therefore true), to the **mathematical universe** (of which geometry is a part) and the **real universe**: the world we live in, seems to respect Euclidean geometry perfectly. A poorly understood abyss had opened up between the **real world** and the worlds constructed out of axioms and precise rules. These worlds we call **geometry**, **arithmetic**, **algebra**, **analysis** and **topology** (see *"The basic achievements of Rigorous Theoretical Logic"*, Appendix 29). These constructions should leave no room for paradoxes, yet logical paradoxes do exist.

Such paradoxes use the same logical structure as mathematical proofs, and Hilbert was concerned about the logical coherence of the mathematical enterprise. His perspective: *"every defined mathematical problem must necessarily be susceptible to an exact solution, either in the form of an actual true answer to the question posed, or through a proof of the impossibility of its solution."*

Hilbert challenged his colleagues to **formalize all mathematical truths in a way that would exclude paradoxical assertions**, like those that arise in ordinary Language and Logic, from Mathematics.

In his address at the Bologna Congress, Hilbert challenged his colleagues to formalize all mathematical truths in a way that would exclude non-demonstrable truths as well as the types of paradoxical assertions that can surface in ordinary Language, like the "liar" and "barber" paradoxes invented by Epimenides and Russell.

Three years after Hilbert's speech at the Bologna Congress, a young ex-physicist, Kurt Gödel, transformed Hilbert's dream in the discovery of the limits existing for an even more rigorous way of thinking, that of Mathematical Logic, which could not be disputed.

The young ex-physicist of the Vienna Circle was not one of those who spoke of the *Principia Mathematica* without having even opened any of the three volumes. Gödel used the *Principia* **like a runway for launching an incredible supersonic airplane** that, once in flight, observed from high above all that which mathematical thinking believed it had **"seen", even if seen only from the ground**.

Gödel was able to demonstrate that even the application of all the tools of Rigorous Logic together left some mathematical truths that still were **not demonstrable**. *Contrary* to what Hilbert had thought, the **abyss** between what is **true** and what **can be demonstrated** is **unbridgeable**. The argument had been liberated from the bindings and dangers **intrinsic to language** by the work of Russell and Whitehead, which **had paved the runway** by reducing **everything to symbols** and **sequences of symbols, and no words**.

Gödel not only read the *Principia*, but also **codified** all the symbols and assertions of the formal logic structure used by Russell and Whitehead, **assigning a single number** to each assertion and every sequence of Arithmetic assertions that could be expressed through the *Principia Mathematica* formalism.

In this formalism there are "elementary logical signs" and "logical variables" linked by "signs". Examples of "elementary logical signs" include the symbol "=", which means "is equal to", and the symbol "∃", which means "exists".

A number is associated with each "**elementary logic sign**". If there are twenty "signs", the first twenty numbers: 1, 2, ... 20, are used to identify them.

There are *three types* of "**logical variables**" that are ordered hierarchically and linked to the precise role of the "variable" within a complex logical expression. The *three types* of variables are: "*numerical*", "*enunciative*" and "*predicative*". The "*numerical*" ones can only have numerical values. The "*enunciative*" ones represent entire formulas. The "*predicative*" ones express properties like "less than", "odd" and "even".

This is the formidable invention of Gödel: all logical expressions and relationships of demonstrability of the Russell and Whitehead treatise can be written by linking these *three types* of "logical variables" with "*elementary* *logical signs*". Gödel codified *"numerical variables"* with *prime numbers*, *"enunciative variables"* with the *squaring* of prime numbers, and *"predicative variables"* with the *cubing* of prime numbers.

The main point is that **each Logical Formula produces** a **number** that **represents it**.

The reader should not despair - we are trying to make comprehensible the **greatest discovery of all times in logical thought**. This discovery had its starting point in the monumental work of **Russell** and **Whitehead**, a work that the great mathematician, famous educator and philosopher **John Kemeny** described as *"a masterpiece discussed by practically all philosophers but read by practically no one"*.

The numeration of Gödel associates one single, unique number to each arithmetic affirmation and each sequence of arithmetic affirmations expressed through the formal logic constructed by Russell and Whitehead in their *Principia Mathematica* using only formulae in order to eliminate the use of common words in the Language of Mathematics. By **making each formula correspond to a number**, Gödel eliminated not only the use of words but also of formulae keeping only numbers in the mathematical Language.

"**Hilbert's Dream**" was to solve all Mathematical problems, and to do this required that all Mathematical problems are described **solely with formulas**, with **no words**. The Principia of **Russell and Whitehead** did precisely this and transformed "Hilbert's Dream" into what was believed to be a **rigorous and monumental edifice**. Gödel proved that the "monumental edifice" was really a runway. **Gödel** in fact transformed the **formulas** of this monumental Work **into numbers**, thereby superseding "Hilbert's Dream" and relegating it to the past.

The realization of Gödel's project invites us to ponder how the real world extends well beyond our human ideas and fantasies. Even the most advanced mathematical structure can only capture **a part** of the picture but **never all of it**. On the other hand, we still need an **analytical instrument** that can help us distinguish the true from the false. This instrument has the origin in what we call the **Reason** that has enabled us to discover three great things: Language, Logic and Science as reported in **Figure 31** (Chapter VI).

APPENDIX 14: *(quoted on pages 81, 83 and 174)* **Wigner, von Neumann and Gödel**

Wigner was schoolmate of von Neumann in Budapest and friend of Gödel in Princeton.

Before the great discovery of Gödel, Johnny von Neumann – during three years (1925–1928) – was trying to rescue mathematics [110]. Von Neumann created a simple and beautiful set of axioms, which were later shown by Kurt Gödel to be exactly what was needed to understand the true nature of Mathematics.

"*Johnny could have discovered what Gödel did*", Wigner told me after a lesson at Erice on his friendship with Gödel.

Photo 29
Wigner in 1973 during a lecture at Erice on his interactions with Gödel.

As reported in Chapter IV, Gödel discovered that the "principle of the **excluded third**" is not valid despite being held to be valid for thousands of years.

Kurt Gödel in Vienna, in 1931, was able to prove that it is impossible to construct a mathematical structure, based on a system of axioms, which is at the same time "consistent" and "complete".

Consistent corresponds to the property that the given set of axioms will never produce a theorem and its negation.

Complete corresponds to the property that every theorem can be proven to be either right or wrong.

The most incredible consequence of Gödel's discovery was the "destruction" of the "*Principia Mathematica*" of Russell and Whitehead, taken to be the solution of all mathematical difficulties based on the elimination of "words" from mathematics. Hilbert's dream to resolve the crisis of mathematics via the elimination of "words" was proved by Gödel to be over. Mathematics, in order to rest on firm logical foundations, needed the proof that every meaningful mathematical statement had to be either "true" or "false".

Gödel discovered the third way: the mathematical "indecidability", one year after Heisenberg discovered the "Uncertainty Principle" in Physics.

Wigner was convinced of the need for profound reflection about the property that distinguishes us from every other form of living matter: Reason.

It is this property which will allow us to understand why in Theoretical Rigorous Logic (Mathematics) there is the "indecidability" while in Experimental Rigorous Logic (Physics) we have only the "Uncertainty Principle", and no "indecidability".

In Chapter III-3.4 the problem of Black Holes is introduced as a new limit to knowledge in Physics.

APPENDIX 15: *(quoted on page 83)* **Immanent and Transcendent**

Our life is an example of existence in the immanentistic and transcendental spheres. With our five senses (sight, hearing, touch, smell and taste) we can interact with the world made of inert matter and matter with life (vegetal and animal). All forms of matter with which we can interact using our five senses represent the Immanent. Those who believe in God need another sphere of existence: the sphere which is totally different from the other one, and is governed by laws which cannot generate reproducible experimental results. This sphere is the Transcendent.

APPENDIX 16: *(quoted on page 83)* **Creativity in mathematical logic: Infinity**

Our intellect has succeeded in inventing infinity as a non-contradictory logical construction. There is not just one but many, many infinites: actually an infinite number. It may seem incredible but it is true. It was discovered by George Cantor (1845–1918).

It took almost a hundred years – and two geniuses of mathematical logic, Kurt Gödel (1906–1978) and Paul Cohen (1934–2007) – to demonstrate that this mathematically rigorous construction is not contradictory in any way.

The infinity is intellect's most formidable invention. Why?

Because *"In the reality surrounding us, everything is finite"* [Readers wishing to know more about this can read my book *L'Infinito*, Il Cigno Galileo Galilei, three editions (1988), Rizzoli-Bur, seven editions (1994–1997), Pratiche Editrice, six editions (1998–2001), Il Saggiatore-NET, two editions (2005–2006), Marco Tropea Editore (2009)].

APPENDIX 17: *(quoted on page 96)* **Why electrons are so important**

Electrons govern processes linked to the functioning of atoms and molecules.

If this particle had not been discovered (the first example of an elementary particle), no one could have conceived how groups consisting of lots of atoms and lots of molecules, which are so important for life, can exist and work so well.

APPENDIX 18: *(quoted on page 102)* **The problem of going from inert matter to living matter**

We are the only form of living matter with the skill to discover **Language**, **Rigorous Theoretical Logic**, better known as **Mathematics**, and **Rigorous Experimental Logic**, better known as **Science**.

This great skill was named **Reason**. It is thanks to Science that we have discovered the seven fundamental components needed to construct the real world as we know it and participate in it. These seven components are: **space** (s), **time** (t), **spin** (σ), **mass** (m), **energy** (E), **gauge charge** (q^G) which generates the Fundamental Forces of Nature and **flavour charge** (q^f) which produces the stability of all fundamental particles, called quarks and leptons. The stability of matter, which every thing is made of, depends on the stability of quarks and leptons. These seven components suffice to describe everything that refers to inert matter.

Life extends beyond the confines of inert matter. When we introduce life into a description of the world, the first mystery to solve is the transition itself from inert matter to living matter. This is now called *"the problem of Minimal Life"*, i.e. how many pieces of inert matter are needed in order to produce the simplest form of a living cell. Hundreds of scientists are working in many Labs (some secret) to solve this problem.

APPENDIX 19: *(quoted on pages 112 and 185)* **The Ten Challenges of our Physics**

Let us briefly indicate these ten challenges. The progress, as far as we can predict now, depends on these ten challenges.

THE TEN CHALLENGES

1) *The Physics of Imaginary Masses: SSB*
2) *Matter-Antimatter Symmetry*
3) *Supersymmetry*
4) *Non-perturbative QCD and the Physics of deconfined colour charges*
5) *Anomalies and Instantons*
6) *Flavour mixing in the quark sector*
7) *Flavour mixing in the leptonic sector*
8) *The problem of the missing mass in the Universe*
9) *The problem of the hierarchy*
10) *The Physics at the Planck Scale, the Gap and the number of expanded dimensions*

APPENDIX 20: *(quoted on page 114)* **Ettore Majorana: Genius and Mystery, what Robert Oppenheimer told me**

After suffering heavy repercussions from his opposition to the development of weapons even stronger than those that destroyed Hiroshima and Nagasaki, Oppenheimer decided to get back to physics by visiting the biggest laboratories at the frontiers of scientific knowledge. This is how he came to CERN, the largest European Laboratory for Subnuclear Physics. At a ceremony organized by Victor Weisskopf, at that time CERN-Director General, for the presentation of the Erice School dedicated to Ettore Majorana, many illustrious physicists participated. I – at the time very young – was entrusted the task of speaking about the Majorana neutrinos. Oppenheimer wanted to voice his appreciation for how the Erice School and the Centre for Scientific Culture had been named. He knew the exceptional contributions Majorana made to physics from the papers he had read. This much, any physicist could do at any time. **What would have remained unknown** is the episode he told me as a testimony of Fermi's exceptional esteem of *"Ettore"*. He recounted the following episode from the time when the Manhattan Project was being carried out. The Project, over the course of less than four years, transformed the scientific discovery of nuclear fission (heavy atomic nuclei can be broken to produce enormous quantities of energy) into a weapon. There were three critical turning points during this Project. During the executive meeting convened to address the first of these crises, Fermi turned to Wigner and said: *"If only Ettore were here."* The Project seemed to have reached a dead end in the second crisis, during which

Fermi exclaimed once more: *"This calls for Ettore!"* Other than the Project Director himself (Oppenheimer), three people were in attendance at these meetings: two scientists (Fermi and Wigner) and a general of the US armed forces. Wigner worked with nuclear forces, like Ettore Majorana. After the "top-secret" meeting, the general asked the great Professor Wigner who this "Ettore" was, and Wigner replied: *"Majorana."* The general asked where Ettore was, so that he could try to bring him to America. Wigner replied: *"Unfortunately, he disappeared in 1938."*

APPENDIX 21: *(quoted on pages 123 and 181)* **A New Manhattan Project for the life and dignity of all people in the world**

The aim of the Manhattan Project was to transform a scientific discovery – the fission of uranium – into a nuclear bomb.

A key factor in its incredible success was the involvement of the cream of European scientists, many of whom had fled from the Nazi-fascist terror pervading Europe in that period.

They worked with great commitment, driven by the terror that the Nazis – who had had a big head start – would beat them in the race to produce the first nuclear bomb in history.

A New Manhattan Project to defend the life and dignity of all people, believers and non-believers alike, would represent the triumph of science for beneficial purposes.

This utopia would be realisable if there was a prevailing culture of love, solidarity and forgiveness between all peoples.

The cover page of the Project and a few statements concerning the Project are reproduced in the following pages.

THE NEW MANHATTAN PROJECT

* * * * * * * *

We live in a culture that ignores the **72 Planetary Emergencies**.
In 1985 the two most powerful Heads of State (USA, USSR)
Reagan and Gorbachev,
said in Geneva that the enemy number one of Peace in the world
are the secret Laboratories; they declared their will to open them,
as proposed by the scientists signatories of the
Erice Statement (see page 137),
for a Science without secrets and without borders.
Unfortunately, even today, the secret Laboratories are closed and
the only solution is to fight the secret in Science at the origins,
with the implementation of the **"New Manhattan Project"**.

* * * * * * * *

* * * * * * * *
The New Manhattan Project ≡ A Project for Mankind
It would be pure utopia if he did not have roots in what has been actually
achieved, starting from the University of Bologna,
in the fight against the Planetary Emergencies.
The scientific-technological results obtained demonstrate
that the problems of the **72 Planetary Emergencies** can be solved.
* * * * * * * *
The Culture of our time, called modern,
ignores the great achievements of Science and brings us back to before 1985,
when the **Future of all Nations** was in danger because of the
60 thousand H-bombs
(each one million tons of TNT equivalent)
accumulated in the arsenals of the two superpowers.
Today, the danger lies in the existence of the **72 Planetary Emergencies**
that threaten the **Future of all Nations** North-South-East-West.
* * * * * * * *
Those who speak of Science
without having ever been able to make a discovery or an invention
are responsible for the
Cultural Hiroshima in which we live and that
the *Three Days* of Bologna
started to bring the focus of all people in the world.
* * * * * * * *
How? Denouncing the existence of as **many as 72 Planetary Emergencies**
and proposing the implementation of a
"New Manhattan Project",
so that the great achievements of Science produce
– quickly – technological inventions
to use in everyday life.
* * * * * * * *
Not letting tens of thousands of years
as it happened with the **Wheel** and the **Fire,**
or one hundred years as it happened with the discovery of the electron.
But within a few years as proved by the **Manhattan Project**.
* * * * * * * *
Enrico Fermi feared, already half a century ago, that
the Political Hiroshima would have been followed by the **Cultural Hiroshima**
in which we are immersed.
* * * * * * * *
The only way to rid the future of the world
from the nightmare of the **72 Planetary Emergencies**
is the realization of the **"New Manhattan Project"**.

APPENDIX 22: *(quoted on page 150)* **The Useless Particles and the Origin of the Universe**

The discovery of the existence of only three columns (also called Families) of particles was made using a supercollider (LEP) at CERN. At LEP billions upon billions of antielectrons collided with billions upon billions of electrons. If we now know the number of families of fundamental particles, it is due to this gigantic machine. But LEP could not have been built if the elusive antiparticle, called the antielectron, predicted by the Dirac equation in 1929, had not been there.

Dirac opened the way towards the search for the "useless" particles and we know today that there are more useless than useful particles. The world familiar to us is made only with particles belonging to the so-called "first Family": the particles in column I of **Figure 46** in Chapter X. The useless particles are those of columns II and III in the same figure.

Nevertheless, it is thanks to the "useless" particles that we are able to trace our existence back to the origin of the universe. In fact, the world opened up by Paul Dirac existed only for a few instants after the Big Bang. At that time, many billions years ago (about 15 billion years), there was a perfect mixture of particles and antiparticles of all Families. Today, after such a long interval of time, the universe has evolved into a world where only particles of the first Family exist.

The roots of the Standard Model are in the Dirac equation. We are all children of this equation. Without it, there would be no particle-physics labs and no Standard Model. Of course – and fortunately for us – there are sound reasons to believe that there is a lot of new physics beyond the Standard Model as illustrated in **Figure 48** of Chapter X.

The recent results on neutrino oscillations and the new values for direct "CP-violation" in K–meson physics are opening the way for new physics beyond the Standard Model. However, none of these discoveries are affected in any way by the gravitational force, which was Einstein's main interest.

The problem of how relevant the consequences of the Dirac equation have been is not new. But it had never been discussed before the early 1970s. At that time, Weisskopf, Eugene Wigner, Bob Wilson and other eminent scientists attended a seminar held in Erice, Sicily, entitled "The Roots of Modern Physics". The conclusion of the seminar was that the effective origin – the real seed of modern physics – was the Dirac equation and the experimental discoveries of the Blackett group.

Three decades later this conclusion has been further corroborated. Imagine modern physics without the possibility of particle-antiparticle symmetry and, consequently, no possibility of annihilation.

As we have repeated several times, there would be no "running" of the gauge couplings and no coupling among different gauge forces (as explained in "*The Virtual Annihilation Phenomena open the horizon for the Grand Unification*", Appendix 27).

Suppose that no one had ever tried to describe the evolution of an electron in Space-Time or thought of the extreme consequences of this work such as the experimental proof of electron and antielectron production and annihilation. We would be nowhere.

It is for these reasons that on the basis of what I know, Blackett and Dirac had the biggest impact in Physics in the twentieth century.

APPENDIX 23: *(quoted on page 174)* **The limits of Human Imagination (Galileo Galilei)**

"What we imagine
must be
either something already seen,
or a combination of things
or parts
of things other times seen,
such as sphinxes,
sirens,
chimeras
centaurs…".

Galileo Galilei, Opere VII, 86.

APPENDIX 24: *(quoted on page 174)* **Beyond the limits of Human Imagination**

All discoveries of scientific origin come from totally unexpected phenomena. No one would have been able to imagine in 1947 the existence of the subnuclear universe [103]. And our ancestors would remain fascinated by the synthesis of our knowledge, called the Standard Model. What is beyond this synthesis no one knows. The only thing we are sure of is that the Standard Model is not the end of our search to decipher the Logic of Nature [9].

Scientific discoveries go beyond the limits of human imagination.

APPENDIX 25: *(quoted on page 181)* **The Roots of LEP and LHC**

The Figure below is the front page of the CERN volume where the roots of LEP and of LHC are reported in all details.

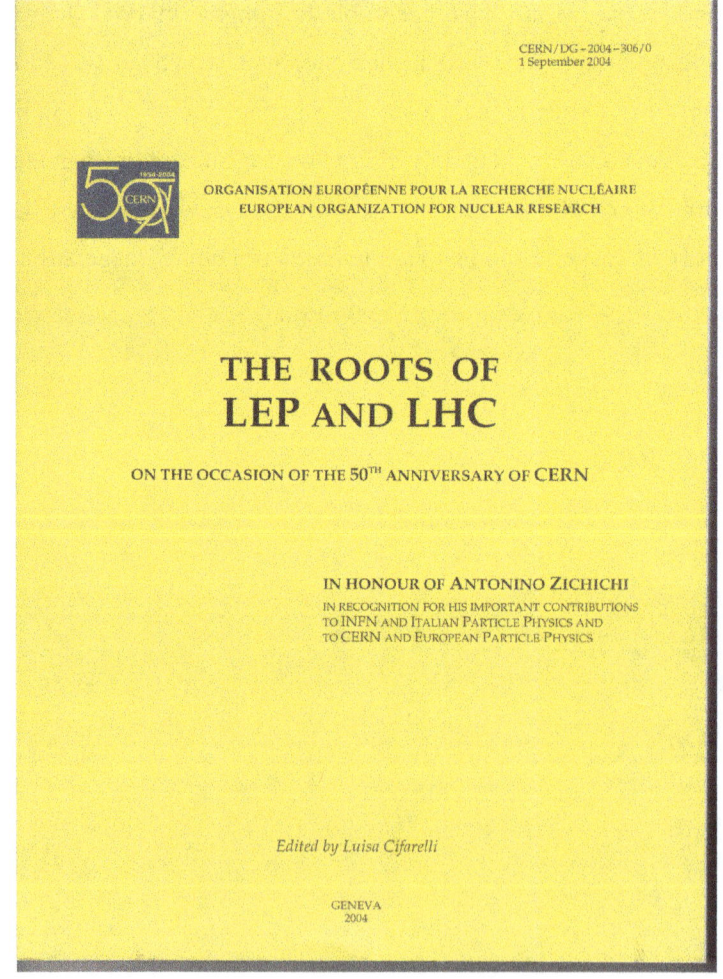

APPENDIX 26: *(quoted on Appendix 1)* **Paul Dirac, Antimatter and 'Virtual Phenomena'**

Dirac developed his equation by studying what happens to an electron as it evolves in space and time.

As a young man, Dirac was fascinated by Lorentz's discovery of the "complex" nature of Space-Time and by the discovery of the first example of an "elementary" particle consisting of nothing else than itself, namely J.J. Thomson's electron.

No one had previously set himself the task of studying the evolution of this simple particle in Space-Time, which Lorentz had found to possess incredibly new characteristics (space and time cannot both be real).

The need for rigour in the description of what happens to the electron led to a turning point which opened up the horizons of Galilean reality to antiparticles and antimatter.

The physics of "virtual phenomena" also stemmed from this need.

The frontiers of modern physics lie with the study of "virtual phenomena".

APPENDIX 27: *(quoted on Appendices 11 and 21)* **The Virtual Annihilation Phenomena open the horizon for the Grand Unification**

Let us try to see how the annihilation establishes a strict connection between all fundamental forces of Nature. This connection between the different forces is the basis for GUT, the Grand Unified Theory of all the fundamental forces (discussed in Chapter III-3).

Think of a photon, which can transform itself into an electron-positron pair and then annihilate to form a photon again. This sequence of processes is governed by Quantum ElectroDynamics (QED).

The same photon can transform **into** a quark-antiquark pair (governed by Quantum ChromoDynamics (QCD) – the theory of the strong force acting between quarks and gluons) **or into** a pair of weak bosons, W^+W^- (governed by Quantum FlavourDynamics (QFD) – the theory of the electroweak force). The different pairs of particle plus antiparticle can later annihilate and form a photon (QED) or a gauge boson of the strong force (QCD) or a gauge boson of the weak force (QFD).

The annihilation allows these three theories (QED, QCD, QFD) to be present in all possible virtual effects. Without these effects, the problem of the renormalization of the gauge forces (with or without spontaneous symmetry breaking) would never have been conceived. And if that problem had not been solved – as it was in the early 1970s by the 1999 Nobel Prize winners Gerard 't Hooft and Martinus Veltman – we would not have had the Standard Model, with its many precise quantitative predictions that have been experimentally validated in labs all over the world.

APPENDIX 28: *(quoted on Appendix 13)* **Hilbert and the German Emperor**

The German Emperor was the only European Head of State to decree the beginning of the twentieth century in 1900 instead of 1901, as all other Kings and Emperors did erroneously. It is not a coincidence that Hilbert was the mathematical advisor to the Emperor. As the reader might recall, the same error repeated itself at the beginning of the twenty-first century. (See A.Z. *L'irresistibile Fascino del Tempo*, il Saggiatore, 2000).

APPENDIX 29: *(quoted on Appendix 13)* **The basic achievements of Rigorous Theoretical Logic**

Arithmetic is the theory of numbers.

Algebra is the theory of relationships among variables (like x, y, z ...), where each variable can be any number.

Analysis is the theory of relationships among functions (like $f = x^3 + y^2 + z$ ...) where each function is constructed with variables. The functions can also be complex functions.

Geometry is the theory of functions – scalar, vector, tensor, etc. - in a determined metric space.

Topology is the theory of relationships among spaces with various properties, with or without various metrics, and with environments. A space that has a metric is, in fact, already constrained/enclosed.

A topological space is a collection of points plus a collection of environments. This makes up a rich branch of Mathematical Logic called the *Proximity theory*.

REFERENCES

Note

In order to help the reader, the reference, where the title of the paper is strictly connected with the topic discussed, is reported directly in the text.

All other references are quoted with a number and reported here.

[1] *The Abdus Salam Dream*
A. Zichichi, CERN/LAA/93-32/a, 21 October 1993. Invited Lecture given at the *Salamfest*, in honour of Abdus Salam, *ICTP*, Trieste, Italy, 8-12 March 1993.

[2] *Galilei, Divine Man*
A. Zichichi, *Italian Physical Society*, two editions (2009–2010).

[3] *Frontiers in Physics*
A. Zichichi, CERN/LAA/93-32, 2 November 1993. Invited Lecture given at the X Anniversary Celebration of the Third World Academy of Sciences, *ICTP*, Trieste, Italy, 1-4 November 1993.

[4] *Abdus Salam Beyond ICTP and the Electroweak Forces*
A. Zichichi, CERN/LAA/95-18, September 1995.

[5] *Examples of the Production of $(K^0, \bar{K}^0)$ and $(K^+, \bar{K}^0)$ Pairs of Heavy Mesons*
W.A. Cooper, H. Filthuth, J.A. Newth, G. Petrucci, R.A. Salmeron and A. Zichichi, *Nuovo Cimento* $\underline{5}$, 1388 (1957).

[6] *Some Lessons from Sixty Years of Theorizing*
M. Gell-Mann, in Proceedings Conference in Honor of Murray Gell-Mann's 80th Birthday, *NTU*, Singapore (2010), *Int. J. Mod. Phys.* $\underline{A25}$ (2010); in the same Proceedings *Murray Gell-Mann and the last frontier of LHC Physics: the QGCW Project*, A. Zichichi; see also *Some Reminiscences of Research Leading to QCD and Beyond,* M. Gell-Mann, in *What We Would like LHC to Give Us*, Proceedings of the 2012–Erice Subnuclear Physics School, Vol. $\underline{50}$, A. Zichichi (ed), *World Scientific* (2014).

[7] *All Possible Symmetries of the S Matrix*
S. Coleman and J. Mandula, *Phys. Rev.* 159, 1251-1256 (1967).

[8] *A Lagrangian Model Invariant Under Supergauge Transformations*
J. Wess and B. Zumino, *Phys. Lett.* B49, 52 (1974); and

Supergauge Transformations in Four Dimensions
J. Wess and B. Zumino, *Nucl. Phys.* B70, 39 (1974).

[9] *Searching for the Superworld*
R.M. Mössbauer, M.J. Duff and S. Ferrara. In honour of A. Zichichi on the Occasion of the Sixth Centenary Celebrations of the University of Turin, Italy. S. Ferrara and R.M. Mössbauer (eds), *World Scientific Series in 20th Century Physics*, Vol. 39 (2007).

[10] *Progress Toward a Theory of Supergravity*
D.Z. Freedman, P. van Nieuwenhuizen and S. Ferrara, *Phys. Rev.* D13, 3214 (1976);

Consistent Supergravity
S. Deser and B. Zumino, *Phys. Lett.* B62, 335 (1976);

Superstrings, M-Theory and Quantum Gravity
M.B. Green, in *Highlights of Subnuclear Physics: 50 Years Later*, Proceedings of the 1997–Erice Subnuclear Physics School, Vol. 35, A. Zichichi (ed), *World Scientific* (1999).

[11] *M Theory (The Theory Formerly Known as Strings)*
M.J. Duff, *Int. J. Mod. Phys.* A11, 5623 (1996) [arXiv:hep-th/9608117]; and

Not the Standard Superstring Review and *From Superspaghetti to Superravioli*
M.J. Duff, in *The Superworld II*, Proceedings of the 1987–Erice Subnuclear Physics School, Vol. 25, A. Zichichi (ed), *Plenum Press*, New York-London (1990) [QCD161:I65:1987]; see also

Supergravity Models
R. Arnowitt, in *From Supersymmetry to the Origin of Space-Time*, Proceedings of the 1993–Erice Subnuclear Physics School, Vol. 31, A. Zichichi (ed), *World Scientific* (1995).

[12] *Where We Stand with the Real Superworld*
A. Zichichi, in *From Superstrings to the Real Superworld*, Proceedings of the 1992–Erice Subnuclear Physics School, Vol. 30, A. Zichichi (ed), *World Scientific* (1993); and *Where can SUSY be?*, in *From Supersymmetry to the Origin of Space-Time*, Proceedings of the 1993–Erice Subnuclear Physics School, Vol. 31, A. Zichichi (ed), *World Scientific* (1995).

[13] *Fine Structure of the Hydrogen Atom by a Microwave Method*
W.E. Lamb and R.C. Retherford, *Phys. Rev.* 72, 241 (1947).

[14] *Some Photographs of the Tracks of Penetrating Radiation*
P.M.S. Blackett and G.P.S. Occhialini, *Proc. Roy. Soc.*, A, Vol. 139, p. 699 (1933).

[15] *Technological Frontiers*
B. Rossi, *Electronic Triggering, Nature*, Vol. 125, p. 636 (1930); *Phys. Zt.*, Vol. 33, p. 304 (1932); *Ac. Lincei*, Vol. 15, p. 735 (1932); and

Photography of Penetrating Corpuscular Radiation
P.M.S. Blackett and G.P.S. Occhialini, *Nature*, Vol. 130, p. 363 (1932).

[16] *Theoretical frontiers*
P.A.M. Dirac, *Proc. Roy. Soc., A*, Vol. 126, p. 360, (1930); *Proc. Camb. Phil. Soc.*, Vol. 26, p. 361 (1930); *Proc. Roy. Soc., A*, Vol. 133, p. 60 (1931); and

Gruppentheorie und Quantenmechanik
H. Weyl, *Hirzel*, Leipzig (1928), 2nd ed., p. 234 (1931).

[17] *Processes Involving Charged Mesons*
C.M.G. Lattes, H. Muirhead, G.P.S. Occhialini and C.F. Powell, *Nature* 159, 694 (1947); and

Observations on the Tracks of Slow Mesons in Photographic Emulsions
C.M.G. Lattes, G.P.S. Occhialini and C.F. Powell, *Nature* 160, 454 (1947).

[18] *Evidence for the Existence of New Unstable Elementary Particles*
G.D. Rochester and C.C. Butler, *Nature* 160, 855 (1947).

[19] [19a] *Search for the Time-like Structure of the Proton*
M. Conversi, T. Massam, Th. Muller and A. Zichichi, *Phys. Lett.* 5, 195 (1963);

[19b] *A Proposal to search for Leptonic Quarks and Heavy Leptons produced by ADONE*
M. Bernardini, D. Bollini, E. Fiorentino, F. Mainardi, T. Massam, L. Monari, F. Palmonari and A. Zichichi, INFN/AE-67/3, 20 March 1967;

[19c] *Limits on the Electromagnetic Production of Heavy Leptons*
V. Alles-Borelli, M. Bernardini, D. Bollini, P.L. Brunini, T. Massam, L. Monari, F. Palmonari and A. Zichichi, *Lettere al Nuovo Cimento* 4, 1156 (1970);

[19d] *CP-Violation in the Renormalizable Theory of Weak Interaction*
M. Kobayashi and T. Maskawa, *Prog. Theor. Phys.* 49, 652 (1973);

[19e] *Limits on the Mass of Heavy Leptons*
M. Bernardini, D. Bollini, P.L. Brunini, E. Fiorentino, T. Massam, L. Monari, F. Palmonari, F. Rimondi and A. Zichichi, *Nuovo Cimento* 17A, 383 (1973);

[19f] *Evidence for Anomalous Lepton Production in e^+-e^- Annihilation*
M.L. Perl et al., *Phys. Rev. Lett.* 35, 1489 (1975). In this paper, none of the above papers [19a, b, c, e] published in 1963, 1967, 1970, 1973 are quoted.

[20] *The Origin of the Third Family – on the XXX Anniversary of the proposal to search for the Third Lepton at ADONE*
C.S. Wu, T.D. Lee, N. Cabibbo, V.F. Weisskopf, S.C.C. Ting, C. Villi, M. Conversi, A. Petermann, B.H. Wiik and G. Wolf; C.S. Wu (ed), a joint publication by University and Academy of Sciences of Bologna, INFN, SIF, Italy (1997), *World Scientific* (1998).

[21] *On the Analysis of τ-Meson data and the Nature of the τ-Meson*
R.H. Dalitz, *Phil. Mag.* 44, 1068 (1953);

Isotopic Spin Changes in τ and θ Decay
R.H. Dalitz, *Proc. Phys. Soc.* A69, 527 (1956);

Present Status of τ-Spin-Parity
R.H. Dalitz, Proceedings of the *Sixth Annual Rochester Conference on High Energy Nuclear Physics*, Interscience Publishers, Inc., New York, 19 (1956); for a detailed record of the events which led to the (θ–τ) puzzle see R.H. Dalitz *Kaon Decays to Pions: the τ–θ Problem* in *History of Original Ideas and Basic Discoveries in Particle Physics*, H.B. Newman and T. Ypsilantis (eds), *Plenum Press*, New York and London, 163 (1994).

[22] *Question of Parity Conservation in Weak Interactions*
T.D. Lee and C.N. Yang, *Phys. Rev.* 104, 254 (1956).

[23] *Experimental Test of Parity Conservation in Beta Decay*
C.S. Wu, E. Ambler, R.W. Hayward, D.D. Hoppes, *Phys. Rev.* 105, 1413 (1957);

Observation of the Failure of Conservation of Parity and Charge Conjugation in Meson Decays: The Magnetic Moment of the Free Muon
R. Garwin, L. Lederman, and M. Weinrich, *Phys. Rev.* 105, 1415 (1957);

Nuclear Emulsion Evidence for Parity Non-Conservation in the Decay Chain $\pi^+\mu^+e^+$
J.J. Friedman and V.L. Telegdi, *Phys. Rev.* 105, 1681 (1957).

[24] *Behavior of Neutral Particles under Charge Conjugation*
M. Gell-Mann and A. Pais, *Phys. Rev.* 97, 1387 (1955).

[25] *Observation of Long-Lived Neutral V Particles*
K. Lande, E.T. Booth, J. Impeduglia, L.M. Lederman, and W. Chinowski, *Phys. Rev.* 103, 1901 (1956).

[26] *On the Conservation Laws for Weak Interactions*
L.D. Landau, *Zh. Éksp. Teor. Fiz.* 32, 405 (1957).

[27] *Remarks on Possible Noninvariance under Time Reversal and Charge Conjugation*
T.D. Lee, R. Oehme, and C.N. Yang, *Phys. Rev.* 106, 340 (1957).

[28] *Evidence for the 2π Decay of the K_2^0 Meson*
J. Christenson, J.W. Cronin, V.L. Fitch, and R. Turlay, *Phys. Rev. Lett.* 13, 138 (1964).

[29] *Anomalous Regeneration of K_1^0 Mesons from K_2^0 Mesons*
R. Adair, W. Chinowsky, R. Crittenden, L.B. Leipuner, B. Musgrave and F.T. Shively
Phys. Rev. 132, 2285 (1963).

[30] M. Gell-Mann, *The Eightfold Way – A Theory of Strong-Interaction Symmetry*, California Institute Technology Synchrotron Lab. Report 20 (1961); *Derivation of Strong Interactions from a Gauge Invariance*, Y. Ne'eman, *Nucl. Phys.* 26, 222 (1961); The experimental confirmation, in 1964, by Samios and collaborators [31] of the existence of the missing member of the baryonic decuplet, Ω^-, appeared to be, at the time, a triumph for the "eightfold way". The choice of the letter Ω, the last in the Greek alphabet, was due to the conviction that this particle was going to be the last ever to be discovered. See also: M. Gell-Mann and Y. Ne'eman, *The Eightfold Way*, *W.A. Benjamin inc.*, New York and Amsterdam (1964).

[31] *Observation of a Hyperon with Strangeness Minus Three*
V.E. Barnes, P.L. Connolly, D.J. Crennell, B.B. Culwick, W.C. Delaney, W.B. Fowler, P.E. Hagerty, E.L. Hart, N. Horwitz, P.V.C. Hough, J.E. Jensen, J.K. Kopp, K.W. Lai, J. Leitner, J.L. Lloyd, G.W. London, T.W. Morris, Y. Oren, R.B. Palmer, A.G. Prodell, D. Radojictc, D.C. Rahm, C.R. Richardson, N.P. Samios, J.R. Sanford, R.P. Shutt, J.R. Smith, D.L. Stonehill, R.C. Strand, A.M. Thorndike, M.S. Webster, W.J. Willis and S.S. Yamamoto, *Phys. Rev. Lett.* 12, 204 (1964).

[32] *A Schematic Model of Baryons and Mesons*
M. Gell-Mann, *Phys. Lett.* 8, 214 (1964);

Fractionally Charged Particles and SU_6
G. Zweig, CERN Report TH 401 (1964); and Erice Lecture 1964, in *Symmetries in Elementary Particle Physics*, Proceedings of the 1964–Erice Subnuclear Physics School, A. Zichichi (ed), *Academic Press*, New York, London (1965).

[33] *Particle Physics for Nuclear Physicists*
H.J. Lipkin, in *Physique Nucléaire*, Les-Houches 1968, C. DeWitt and V. Gillet (eds), *Gordon and Breach*, New York, 585 (1969);

Triality, Exotics and the Dynamical Basis of the Quark Model
H.J. Lipkin, *Phys. Lett.* 45B, 267 (1973);

A Systematics of Hadrons in Subnuclear Physics
Y. Nambu, in *Preludes in Theoretical Physics*, A. de-Shalit, H. Feshbach and L. Van Hove (eds), *North Holland Pub. Comp.*, Amsterdam, 133 (1966).

[34] *Spin and Unitary-Spin Independence in a Paraquark Model of Baryons and Mesons*
O.W. Greenberg, *Phys. Rev. Lett.* 13, 598 (1964).

[35] *Three-Triplet Model with Double SU(3) Symmetry*
M.Y. Han and Y. Nambu, *Phys. Rev.* 139B, 1006 (1965).

[36] *Advantages of the Color Octet Gluon Picture*
H. Fritzsch, M. Gell-Mann and H. Leutwyler, *Phys. Lett.* 47B, 365 (1973).

[37] *Can We Make Sense Out of "Quantum ChromoDynamics"?*
G. 't Hooft, in *The Whys of Subnuclear Physics*, Proceedings of the 1977–Erice Subnuclear Physics School, Vol. 15, A. Zichichi (ed), *Plenum Press*, New York and London, 943 (1978).

[38] For a complete set of references concerning this topic see *The Creation of Quantum ChromoDynamics and the Effective Energy*, V.N. Gribov, G. 't Hooft, G. Veneziano and V.F. Weisskopf; L.N. Lipatov (ed), Academy of Sciences and University of Bologna, INFN, SIF, Italy, published by *World Scientific* (1998).

[39] *Evidence of the Same Multiparticle Production Mechanism in p-p Collisions as in e^+e^- Annihilation*
M. Basile, G. Cara Romeo, L. Cifarelli, A. Contin, G. D'Alì, P. Di Cesare, B. Esposito, P. Giusti, T. Massam, F. Palmonari, G. Sartorelli, G. Valenti and A. Zichichi, *Phys. Lett.* 92B, 367 (1980).

[40] *Quark Search at the ISR*
T. Massam and A. Zichichi, *CERN preprint*, June 1968;

Search for Fractionally Charged Particles Produced in Proton-Proton Collisions at the Highest ISR Energy
M. Basile, G. Cara Romeo, L. Cifarelli, P. Giusti, T. Massam, F. Palmonari, G. Valenti and A. Zichichi, *Nuovo Cimento* 40A, 41 (1977); and

A Search for Quarks in the CERN SPS Neutrino Beam
M. Basile, G. Cara Romeo, L. Cifarelli, A. Contin, G. D'Alì, P. Giusti, T. Massam, F. Palmonari, G. Sartorelli, G. Valenti and A. Zichichi, *Nuovo Cimento* 45A, 281 (1978).

[41] *Gauge Theories with Unified, Weak, Electromagnetic and Strong Interactions*
G. 't Hooft, in EPS International Conference on *High Energy Physics*, Palermo, 23-28 June 1975, A. Zichichi (ed), *Editrice Compositori*, Bologna, Italy, 1225 (1976).

[42] *Cloud Chamber Observations of Cosmic Rays at 4300 Meters Elevation and Near Sea Level*
C.D. Anderson and S.H. Neddermeyer, *Phys. Rev.* 50, 263 (1936).

[43] *On the Disintegration of Negative Mesons*
M. Conversi, E. Pancini and O. Piccioni, *Phys. Rev.* **71**, 209 (1947).

[44] *Sui Mesoni dei Raggi Cosmici*
G. Puppi, *Nuovo Cimento* **5**, 587 (1948). In this paper G. Puppi suggests that all Fermi processes could be described by the same coupling. In fact the decay rates of three different processes (π decay), (μ capture) and (μ decay) were found to be "approximately" the same. This is the origin of the Puppi triangle

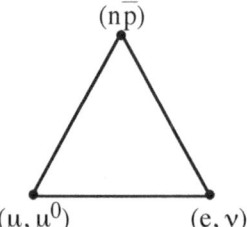

where the three vertices allow, through their couplings, to describe all known weak processes at that time. Note that Puppi distinguishes the neutral counter part of the muon, μ_0, (now known as ν_μ) from the neutral counter part of the electron, ν, (now called ν_e). The existence of a second neutrino, i.e. $\nu_\mu \neq \nu_e$, was established in 1962 by L.M. Lederman, M. Schwartz, J. Steinberger and Collaborators at BNL [45]. Other physicists have contributed to the universality of the Fermi interactions [46–48].

[45] *Observations of High-Energy Neutrino Reactions and the Existence of Two Kinds of Neutrinos*
G. Danby, J.-M. Gaillard, K. Goulianos, L.M. Lederman, N. Mistry, M. Schwartz and J. Steinberger, *Phys. Rev. Lett.* **9**, 36 (1962).

[46] *Mesons and Nucleons*
O. Klein, *Nature* **161**, 897 (1948).

[47] *Interaction of Mesons with Nucleons and Light Particles*
T.D. Lee, M. Rosenbluth and C.N. Yang, *Phys. Rev.* **75**, 905 (1949).

[48] *Energy Spectrum of Electrons from Meson Decay*
J. Tiomno and J.A. Wheeler, *Rev. Mod. Phys.* **21**, 144 (1949).

[49] *The Anomalous Magnetic Moment of the Muon*
G. Charpak, F. Farley, R.L. Garwin, T. Muller, J.C. Sens, V.L. Telegdi, C.M. York and A. Zichichi, Proceedings of the International Conference on *High-Energy Physics*, Rochester, NY, USA, 25 August-1 September 1960, Univ. Rochester, 778 (1960).

[50] *Measurement of the Anomalous Magnetic Moment of the Muon*
G. Charpak, F.J. Farley, R.L. Garwin, T. Muller, J.C. Sens, V.L. Telegdi and A. Zichichi, *Phys. Rev. Lett.* **6**, 128 (1961).

[51] *A New Measurement of the Anomalous Magnetic Moment of the Muon*
G. Charpak, F.J. Farley, R.L. Garwin, T. Muller, J.C. Sens and A. Zichichi, *Phys. Lett.* **1**, 16 (1962).

[52] *(g–2) and Its Consequences*
G. Charpak, F.J. Farley, R.L. Garwin, T. Muller, J.C. Sens and A. Zichichi
Proceedings of the International Conference on *High-Energy Physics*, Geneva, Switzerland, 4-11 July 1962, 476, CERN, Geneva, 1962.

[53] *The Anomalous Magnetic Moment of the Muon*
G. Charpak, F.J. Farley, R.L. Garwin, Th. Muller, J.C. Sens and A. Zichichi, *Nuovo Cimento* 37, 1241 (1965).

[54] *A Measurement of the μ^+ Lifetime*
F.J. Farley, T. Massam, T. Muller and A. Zichichi
Proceedings of the International Conference on *High-Energy Physics*, Geneva, Switzerland, 4-11 July 1962, 415, CERN, Geneva, 1962; and

CERN Work on Weak Interactions
A. Zichichi, in the February 1964 Meeting of the Royal Society. *Proc. Roy. Soc.* A285, 175 (1965).

[55] *A Measurement of the e^+ Polarization in Muon Decay: the e^+ Annihilation Method*
A. Buhler, N. Cabibbo, M. Fidecaro, T. Massam, Th. Muller, M. Schneegans and A. Zichichi, *Phys. Lett.* 7, 368 (1963).

[56] *Lepton Physics at CERN and Frascati*
N. Cabibbo (ed), *World Scientific Series in 20th Century Physics*, Vol. 8 (1995).

[57] *Evidence for a New Decay Mode of the X^0-Meson: $X^0 \to 2\gamma$*
D. Bollini, A. Buhler-Broglin, P. Dalpiaz, T. Massam, F. Navach, F.L. Navarria, M.A. Schneegans and A. Zichichi, *Nuovo Cimento* 58A, 289 (1968).

[58] *Measurement of the Branching Ratio $\Gamma(X^0 \to \gamma\gamma) / \Gamma(X^0 \to TOTAL)$*
P. Dalpiaz, P.L. Frabetti, T. Massam, F.L. Navarria and A. Zichichi, *Phys. Lett.* 42B, 377 (1972).

[59] *The Decay Mode $\omega \to e^+e^-$ and a Direct Determination of the $\omega - \phi$ Mixing Angle*
D. Bollini, A. Buhler-Broglin, P. Dalpiaz, T. Massam, F. Navach, F.L. Navarria, M.A. Schneegans and A. Zichichi, *Nuovo Cimento* 57A, 404 (1968); see also

Observation of the Rare Decay Mode of the ϕ-Meson: $\phi \to e^+e^-$
D. Bollini, A. Buhler-Broglin, P. Dalpiaz, T. Massam, F. Navach, F.L. Navarria, M.A. Schneegans and A. Zichichi, *Nuovo Cimento* 56A, 1173 (1968).

[60] *Radiative Decays of Mesons*
R.H. Dalitz and A. Zichichi (eds), a joint publication by University and Academy of Sciences of Bologna, Italy (2000);

The Basic SU(3) Mixing: $\omega_8 \longleftrightarrow \omega_1$
A. Zichichi, in *Evolution of Particle Physics*, Academic Press Inc., New York and London, 299 (1970).

[61] *A PCAC Puzzle: $\pi^0 \to \gamma\gamma$ in the σ-Model*
J.S. Bell and R. Jackiw, *Nuovo Cimento* A60, 47 (1969).

[62] *Axial-Vector Vertex in Spinor Electrodynamics*
S.L. Adler, *Phys. Rev.* 177, 2426 (1969).

[63] *Computation of the Quantum Effects due to a four-Dimensional Pseudoparticle*
G. 't Hooft, *Phys. Rev.* D14, 3432 (1976); and err. *Phys. Rev.* D18, 2199 (1978); and

How Instantons Solve the U(1) Problem
G. 't Hooft, *Phys. Rept.* 142, 357 (1986).

[64] *Evidence for η' Leading*
L. Cifarelli, T. Massam, D. Migani and A. Zichichi, in *Highlights of Subnuclear Physics: 50 Years Later*, Proceedings of the 1997–Erice Subnuclear Physics School, Vol. 35, A. Zichichi (ed), *World Scientific* (1999). See also D. Migani *Thesis*, Bologna University, Italy (July 1997).

[65] *A Study of the Various Approaches to M_{GUT} and a_{GUT}*
F. Anselmo, L. Cifarelli and A. Zichichi, *Nuovo Cimento* 105A, 1335 (1992).

[66] *The Gran Sasso Project*
A. Zichichi, INFN/AE-82/1, 28 February 1982;

The Gran Sasso Project
A. Zichichi, Proceedings of the Workshop on *Science Underground*, Los Alamos, NM, USA, 27 September-1 October 1982, *AIP*, New York, 52 (1983);

Perspectives of Underground Physics: The Gran Sasso Project
A. Zichichi, Invited Plenary Lecture at the Symposium on *Present Trends, Concepts and Instruments of Particle Physics*, in honour of Marcello Conversi's 70th birthday, Rome, Italy, 3-4 November 1987, G. Baroni, L. Maiani and G. Salvini (eds), Conference Proceedings, SIF, Bologna, Italy, Vol. 15, 107 (1988).

[67] *The Evolution of Gaugino Masses and the SUSY Threshold*
F. Anselmo, L. Cifarelli, A. Petermann and A. Zichichi, *Nuovo Cimento* 105A, 581 (1992).

[68] *Search for Supersymmetric Particles using Acoplanar Charged Particle Pairs from Z^0 decays*
ALEPH Collab., D. Decamp et al., *Phys. Lett.* B236, 86 (1990).

[69] *Search for Neutral Higgs Bosons from Supersymmetry in Z decays*
ALEPH Collab., D. Decamp et al., *Phys. Lett.* B237, 291 (1990).

[70] *Search for Neutralino Production in Z decays*
ALEPH Collab., D. Decamp et al., *Phys. Lett.* B244, 541 (1990).

[71] *Search for the Neutral Higgs Bosons of the MSSM and other two Doublet Models*
ALEPH Collab., D. Decamp et al., *Phys. Lett.* B265, 475 (1991).

[72] *Search for Heavy Charged Scalars in Z^0 decays*
DELPHI Collab., P. Abreu et al., *Phys. Lett.* B241, 449 (1990).

[73] *Search for Pair Production of Neutral Higgs Bosons in Z^0 decays*
DELPHI Collab., P. Abreu et al., *Phys. Lett.* B245, 276 (1990).

[74] *Search for Scalar Quarks in Z^0 decays*
DELPHI Collab., P. Abreu et al., *Phys. Lett.* B247, 148 (1990).

[75] *A Search for Sleptons and Gauginos in Z^0 Decays*
DELPHI Collab., P. Abreu et al., *Phys. Lett.* B247, 157 (1990).

[76] *Mass Limits for Scalar Muons, Scalar Electrons and Winos from e^+e^- Collisions near $S^{**}(1/2) = 91-GeV$*
L3 Collab., B. Adeva et al., *Phys. Lett.* B233, 530 (1989).

[77] *Search for the Neutral Higgs Bosons of the Minimal Supersymmetric Standard Model from Z^0 Decays*
L3 Collab., B. Adeva et al., *Phys. Lett.* B251, 311 (1990).

[78] *Search for the Charged Higgs Boson in Z^0 decay*
L3 Collab., B. Adeva et al., *Phys. Lett.* B252, 511 (1990).

[79] *A Search for Acoplanar Pairs of Leptons or Jets in Z^0 decays: Mass Limits on Supersymmetric Particles*
OPAL Collab., M.Z. Akrawy et al., *Phys. Lett.* B240, 261 (1990).

[80] *A Search for Technipions and Charged Higgs Bosons at LEP*
OPAL Collab., M.Z. Akrawy et al., *Phys. Lett.* B242, 299 (1990).

[81] *A Direct Search for Neutralino Production at LEP*
OPAL Collab., M.Z. Akrawy et al., *Phys. Lett.* B248, 211 (1990); P.D. Acton et al., preprint CERN-PPE/91-115, 22 July 1991.

[82] *Searches for Supersymmetric Particles Produced in Z Boson decay*
MARK II Collab., T. Barklow et al., *Phys. Rev. Lett.* 64, 2984 (1990).

[83] *Searches for New Particles at LEP*
M. Davier, LP-HEP 91 Conference, Geneva, CH, Preprint LAL 91-48, December 1991.

[84] *A Detailed Comparison of LEP Data with the Predictions of the Minimal Supersymmetric SU(5) GUT*
J.R. Ellis, S. Kelley, D.V. Nanopoulos, preprint CERN-TH/6140-91, *Nucl. Phys.* B373, 55 (1992).

[85] *A Method for Trapping Muons in Magnetic Fields, and its Application to a Redetermination of the EDM of the Muon*
G. Charpak, L.M. Lederman, J.C. Sens and A. Zichichi, *Nuovo Cimento* 17, 288 (1960); see also references [49–53].

[86] *Present Status of Strong Electromagnetic and Weak Interaction*
R.P. Feynman in *Symmetries in Elementary Particle Physics*, Proceedings of the 1964 Erice Subnuclear Physics School, A. Zichichi (ed), *Academic Press*, New York, London (1965).

[87] *A High–Intensity, Partially Separated, Beam of Antiprotons and K–Mesons*
G. Brautti, G. Fidecaro, T. Massam, M. Morpurgo, Th. Muller, G. Petrucci, E. Rocco, P. Schiavon, M. Schneegans and A. Zichichi, *Nuovo Cimento* 38, 1861 (1965);

A High Intensity Enriched Beam of Kaons and Antiprotons
G. Brautti, G. Fidecaro, T. Massam, M. Morpurgo, Th. Muller, G. Petrucci, E. Rocco, P. Schiavon, M. Schneegans and A. Zichichi, Proceedings of the International Conference on *High-Energy Physics*, Dubna, USSR, 5-15 August 1964, *Atomizdat*, Moscow, Vol. II, 556 (1966).

[88] *A Telescope to Identify Electrons in the Presence of Pion Background*
T. Massam, Th. Muller and A. Zichichi, CERN Report 63-25, 27 June 1963.

A New Electron Detector with High Rejection Power Against Pions
T. Massam, Th. Muller, M. Schneegans and A. Zichichi, *Nuovo Cimento* 39, 464 (1965).

[89] *Range Measurements for Muons in the GeV Region*
A. Buhler, T. Massam, Th. Muller and A. Zichichi, CERN Report 64-31, 24 June 1964.

Range Measurements for Muons in the GeV Region
A. Buhler, T. Massam, Th. Muller and A. Zichichi, *Nuovo Cimento* 35, 759 (1965).

[90] *Un Grand Détecteur E.M. à Haute Réjection des Pions*
D. Bollini, A. Buhler-Broglin, P. Dalpiaz, T. Massam, F. Navach, F.L. Navarria, M.A. Schneegans et A. Zichichi, *Revue de Physique Appliquée* 4, 108 (1969);

A Large Electromagnetic Shower Detector with High Rejection Power Against Pions
M. Basile, J. Berbiers, D. Bollini, A. Buhler-Broglin, P. Dalpiaz, P.L. Frabetti, T. Massam, F. Navach, F.L. Navarria, M.A. Schneegans and A. Zichichi, *Nuclear Instruments and Methods* 101, 433 (1972).

[91] *Über das Gravitationsfeld eines Massenpunktes nach der Einsteinschen Theorie*
K. Schwarzschild, *Sitzungsberichte der Königlich Preussischen Akademie der Wissenschaften* 7: 189-196 (1916).

[92] *Dynamics of a Lattice Model Universe by the Schwarzschild-Cell Method*
R.W. Lindquist and J.A. Wheeler, *Rev. Mod. Phys.* 29, 432 (1957).

[93] *Bits, Quanta and Meaning*
J.A. Wheeler, in *Theoretical Physics Meeting: Commemorative Volume on the Occasion of Eduardo Caianiello's Sixtieth Birthday*, A. Giovanni, M. Marinaro, F. Mancini and A. Rimini (eds), *Edizioni Scientifici Italiani*, Naples, pp. 121-134 (1984).

[94] *The Lesson of the Black Hole*
J.A. Wheeler, *Proc. Am. Phil. Soc.* 125, 25 (1981).

[95] *Delayed-Choice Experiments and the Bohr-Einstein Dialogue*
J.A. Wheeler, *The American Philosophical Society and the Royal Society*, pp. 25-37, papers read at the Meeting, June 5, 1980, *American Philosophical Society*, Philadelphia (1981).

[96] *General Relativity and John Archibald Wheeler*
I. Ciufolini and R.A. Matzner (eds), *Springer* (2010).

[97] *The Computer and the Universe*
J.A. Wheeler, *Int. J. Theor. Phys.* 21, Nos. 6/7 (1982).

[98] *It is as if we come from a Black–Hole and we are in a Black–Hole*
A. Zichichi, to be published.

[99] *Evidence for a New 2π Resonance at 700 MeV*
M. Feldman, W. Frati, J. Halpern, A. Kanofsky, M. Nussbaum, S. Richert, P. Yamin, A. Choudry, S. Devons, and J. Grunhaus, *Phys. Rev. Lett.* 14, 869 (1965).

[100] *Experimental Evidence against the Existence of the S^0–meson*
A. Buhler-Broglin, P. Dalpiaz, T. Massam, F.L. Navarria, M. Schneegans, F. Zetti and A. Zichichi, *Nuovo Cimento* 49A, 183 (1967).

[101] *Experimental Observation of Antideuteron Production*
T. Massam, Th. Muller, B. Righini, M. Schneegans and A. Zichichi, *Nuovo Cimento* 39, 10 (1965).

[102] *La Normalisation des Constantes dans la Théorie des Quanta*
A. Petermann and E. Stueckelberg, *Helv. Phys. Acta* 26, 499 (1953).

[103] A. Zichichi, *Subnuclear Physics: The first 50 Years: Highlights from Erice to ELN*, O. Barnabei, P. Pupillo, F. Roversi Monaco (eds), a joint publication by University and Academy of Sciences of Bologna, Italy (1998); *World Scientific Series in 20th Century Physics*, Vol. 24 (2000).

The References [104–107] refer to the various occasions where I have presented papers on highly specialized topics and discussed the connection of these topics with Complexity. Each title on the upper part refers to the connection with Complexity while the specialized topic is reported in the detailed references.

[104] *Complexity at the Fundamental Level*
A. Zichichi
presented at:
- International Conference on 'Quantum [un]speakables' in Commemoration of John S. Bell, International Erwin Schrödinger Institut (ESI), Universität Wien (Austria), November 2000, *'John Bell and the Ten Challenges of Subnuclear Physics'*.

- 40th Course of the International School of Subnuclear Physics, Erice (Italy), September 2002, *'Language Logic and Science'*.
- 31st, 32nd and 33th Course of the International School of Solid State Physics, Erice (Italy), July 2004, *'Complexity at the Elementary Level'*.
- 42nd International School of Subnuclear Physics, Erice (Italy), August - September 2004, *'Complexity at the Elementary Level'*.
- Trinity College, Dublin (Ireland), February 2005, *'Complexity at the Elementary Level'*.
- Department of Physics, University of Padova (Italy), March 2005, *'Complexity at the Elementary Level'*.
- 43th Course of the International School of Subnuclear Physics, Erice (Italy), September 2005, *'Complexity at the Elementary Level'*.
- Italian Physics Society (SIF) XCI Annual National Congress, University of Catania (Italy), September 2005, *'Complexity at the Elementary Level'*.
- DESY, Hamburg, November 2005, *'Complexity at the Fundamental Level'*.
- 44th Course of the International School of Subnuclear Physics, Erice (Italy), September 2006, *'Complexity at the Fundamental Level'*.

[105] *The Logic of Nature and Complexity*
A. Zichichi
presented at:
- Pontificia Academia Scientiarum, The Vatican, Rome (Italy), November 2002, *'Scientific Culture and the Ten Statements of John Paul II'; 'Elements of Rigour in the Theory of Evolution'*.
- The joint Session of:
  6th Course of the International School of Biological Magnetic Resonance; Erice (Italy), July 2003, *'Language Logic and Science'*.
- 2nd Workshop on Science and Religion of the Advanced School of History of Physics; Erice (Italy), July 2003, *'Language Logic and Science'*.
- 10th Workshop of the International School of Liquid Crystals; Erice (Italy), July 2003, *'Language Logic and Science'*.
- International School on Complexity, 1st Workshop on Minimal Life, Erice (Italy), December 2004, *'Evolution and Complexity at the Elementary Level'*.

[106] *Complexity and New Physics*
A. Zichichi
presented at:
- INFN-Alice Meeting, University of Catania (Italy), January 2005, *'Complexity at the Elementary Level'*.
- INFN Eloisatron Project 'The 1st Physics ALICE Week', Erice (Italy), December 2005, *'Complexity and New Physics with ALICE'*.
- 50th Anniversary of INFN Bologna - ALICE Week, Bologna (Italy), June 2006, *'Complexity at the Fundamental Level'*.

[107] *Complexity and Planetary Emergencies*
A. Zichichi
presented at:
- 27th Sessions of the International Seminars on Planetary Emergencies, Erice (Italy), August 2002, *'Language, Logic and Science'*.
- 28th Sessions of the International Seminars on Planetary Emergencies, Erice (Italy), August 2003, *'Language Logic and Science, Evolution and Planetary Emergencies'*.

- 36th Sessions of the International Seminars on Planetary Emergencies, Erice (Italy), August 2006, *'Complexity and Planetary Emergencies'*.

[108] *Complexity exists at the Fundamental Level*
A. Zichichi in *How and Where to Go Beyond the Standard Model*, Proceedings of the 2004–Erice Subnuclear Physics School, Vol. 42, A. Zichichi (ed), *World Scientific*, pp. 251-333 (2007).

[109] *The Logic of Nature, Complexity and New Physics: From Quark-Gluon Plasma to Superstrings, Quantum Gravity and Beyond*
Proceedings of the 2006–Erice Subnuclear Physics School, Vol. 44, A. Zichichi (ed), *World Scientific* (2008).

[110] *Eine Axiomatisierung der Mengenlehre*
J. von Neumann, *J. Math.* 154, 219-240 (1925);

Zur Hilbertschen Beweistheorie
J. von Neumann, *Math Zeitschr.* 26, 1-46 (1927);

Die Axiomatisierung der Mengenlehre
J. von Neumann, *Math Zeitschr.* 27, 669-752 (1928).

[111] *The End of a Myth: High–P_T Physics*
M. Basile, J. Berbiers, G. Cara Romeo, L. Cifarelli, A. Contin, G. D'Alì, C. Del Papa, P. Giusti, T. Massam, R. Nania, F. Palmonari, G. Sartorelli, M. Spinetti, G. Susinno, L. Votano and A. Zichichi, *Opening Lecture* in *Quarks, Leptons, and their Constituents*, Proceedings of the 1984–Erice Subnuclear Physics School, Vol. 22, A. Zichichi (ed), *Plenum Press*, New York-London, 1 (1988).

[112] *Interaction of Elementary Particles*
H. Yukawa, Part I, *Proc. Physico-Math. Soc. Japan* 17, 48 (1935); and

Models and Methods in the Meson Theory
H. Yukawa, *Reviews of Modern Physics* 21, 474 (1949).

[113] *Note on the Nature of Cosmic Ray Particles*
S.H. Neddermeyer and C.D. Anderson, *Phys. Rev.* 51, 884 (1937).

[114] *New Evidence for the Existence of a Particle of Mass Intermediate between the Proton and Electron*
J.C. Street and E.C. Stevenson, *Phys. Rev. (L)* 52, 1003 (1937).

[115] *On the Nature of Cosmic Ray Particles*
Y. Nishina, M. Takeuchi and T. Ichimiya, *Phys. Rev.* 52, 1198 (1937).

[116] *Cosmic-Ray Particles of Intermediate Mass*
S.H. Neddermeyer and C.D. Anderson, *Phys. Rev.* 54, 88 (1938).

[117] *On the Mass of the Mesotron*
Y. Nishina, M. Takeuchi and T. Ichimiya, *Phys. Rev. (L)* 55, 585 (1939).

[118] *The Decay of Megative Mesotrons in Matter*
E. Fermi, E. Teller and V.F. Weisskopf, *Phys. Rev.* 71, 314 (1947).

ACRONYMS

AdA: Anello di Accumulazione (A machine where electrons are accelerated and accumulated).
ADONE: Big AdA.
AFB: Anderson-Feynman-Beethoven-type phenomena.
ALICE: A Large Ion Collider Experiment.
ATLAS: A Toroidal Lhc ApparatuS.
c: The velocity of light.
C: Charge Operator which changes a charge with its anticharge.
CERN: European Centre for Nuclear and Subnuclear Research.
CMS: Compact Muon Solenoid.
CP: Two Operators in action: Charge and Parity. The result is to change a charge with its anticharge and to change right with left.
CPT: Change charge with anticharge, left and right plus the Time arrow.
DESY: Deutsches Elektronen-Synchrotron: the electron accelerator built in Hamburg (Germany) and the name of the Laboratory.
DIS: Deep Inelastic Scattering.
EBUS: Evolution of our Basic Understanding of the laws governing the world in its Structure.
EGM: Evolution of Gaugino Masses.
E_{GUT}: Energy Level where all Fundamental Forces of Nature converge.
ELN: Euroasiatic Long Intersecting Storage Accelerator.

251

EM: ElectroMagnetism.
EMFCSC: Ettore Majorana Foundation and Centre for Scientific Culture.
E_{Planck}: Planck Energy.
EPS: European Physical Society.
E_{SU}: Energy level of the String Theory.
EWRL: Evolution of the World in its Real Life.
GAP: Energy Interval Between E_{Planck} and E_{GUT}.
GeV One billion electron-Volt of Energy.
GUT: Grand Unified Theory.
HERA: Hadron-Electron Ring Accelerator (at DESY, in Germany).
HL: Heavy Lepton.
HZ: Hertz = one cycle per second.
ICSC-WL: International Centre for Scientific Culture Word Laboratory.
ICT: Information and Communication Technology.
ICTP: International Centre for Theoretical Physics.
INFN: Istituto Nazionale di Fisica Nucleare: the largest Italian Organization in the Field of Nuclear and Subnuclear Physics.
ISR: Intersecting Storage Rings. The first proton collider ever built. Its maximum energy was 31 GeV per proton beam. $E_{total} = 62$ GeV.
ITU: International Telecommunication Union.
LAA: Large Asymmetry Analyser (A project to study the invention of new Technologies for the new Colliders at CERN).
LEP: Large Electron Positron Collider.
LHC: Large Hadron Collider.
LOY: T.D. Lee, R. Oehme, C.N. Yang.
MARK-I: The name of an experiment at DESY.
MeV: One million electron-Volt of Energy.
MHZ: MegaHertz (one million Hertz): the unit used to measure the energy level of the hydrogen atom.
NBC: Non-Bubble-Chamber technology.
P: Parity Operator which changes left with right and viceversa.

PAPLEP:	Proton AntiProton Annihilation into LEpton Pairs.
PC:	Identical to CP.
PETRA:	The (e^+e^-) collider built at DESY, in Germany.
PeV:	The Energy of one million of a billion of electron-Volt.
PMP:	Permanent Monitoring Panel.
PS:	Proton Synchrotron.
Q:	Charge.
QCD:	Quantum ChromoDynamics.
QED:	Quantum ElectroDynamics.
QFD:	Quantum FlavourDynamics.
R&D:	Research and Development.
RGEs:	Renormalization Group Equations.
RQST:	Relativistic Quantum String Theory.
S:	Scalar.
SC:	SyncroCyclotron.
SLAC:	Stanford Linear Accelerator Center.
SM&B:	Standard Model and Beyond.
SPEAR:	The Stanford Positron Electron Accelerating Ring.
SPS:	Super Proton Synchrotron.
SSB:	Spontaneous Symmetry Breaking.
SU(3):	A mathematical Symmetry based on 3 quantities.
SUSY:	Supersymmetry.
T:	The operator to change the arrow of Time. From past to future to its opposite: from future to past. Time Reversal Invariance corresponds to all Physics phenomena remaining the same, when the Time arrow is reversed.
TeV:	The Energy of one thousand of a billion of electron-Volt.
TOF:	Time Of Flight.
UEEC:	Unexpected Events with Enormous Consequences.
WFS:	World Federation of Scientists.
WSIS:	World Summit on the Information Society.

INDEX OF NAMES

Adair, Robert, 30.
Adams, Sir John Bertram (1920–1984), 182.
Alderdice, Lord John, 139.
Anderson, Carl David (1905–1991), 41, 143, 147, 181, 206.
Anselmo, Franco, 56.
Arber, Werner, 139.
Archimedes of Syracuse (c. 287 BC – c. 212 BC), 154, 155, 156.
Ashkin, J., 166.
Astbury, Joseph Peter (1916–1987), 79.
Azimov, Yakov, 162.

Balkanski, Minko, 168.
Bakker, Cornelius Jan (1904–1960), 182.
Bardeen, John (1908-1991), 196.
Barnabei, Ottavio (1925–2009), 146.
Basile, Maurizio, 35.
Becquerel, Antoine Henri (1852–1908), 95, 100.
Bednorz, Johannes Georg, 196.
Beethoven, Ludwig van (1770–1827), 143, 144.
Bell, John Stewart (1928–1990), 109, 164, 183.
Bergmann, Peter Gabriel (1915–2002), 180.
Berman, Sam M., 166.
Bernardini, M., 74.
Bernardini, Gilberto (1906–1995), 166.
Bertolucci, Sergio, 98.
Bethe, Hans Albrech (1906–1995), 179, 209.
Biedenharn, L.C., 166.
Bignami, Enrico, 114.
Blackett, Lady Eva Costanza Bernardino Bayon (1899–1986), 85.
Blackett, Lord Patrick Maynard Stuart (1897–1974), 1, 2, 3, 5, 6, 8, 9, 10, 11, 12, 13, 14, 16, 18, 20, 21, 22, 24, 25, 27, 28, 30, 40, 44, 46, 47, 48, 55, 60, 61, 68, 71, 77, 78, 79, 81, 84, 85, 87, 88, 90, 91, 93, 94, 95, 96, 97, 99, 102, 108, 109, 111, 113, 124, 125, 126, 134, 137, 138, 139, 140, 147, 149, 156, 159, 171, 172, 175, 178, 181, 182, 183, 184, 185, 186, 208, 209, 210.

Blackman, (Morris) Moses (1908–1983), 16.
Block, M.M., 166.
Bohr, Niels David (1885–1962), 3, 13, 20, 21, 176, 177.
Bollini, Dante, 74.
Bolyai, János (1802–1860), 214.
Boschi, Enzo, 168.
Bose, Satyendra Nath (1894–1974), 126.
Brout, Robert (1928–2011), 98.
Brunelli, Bruno, 14.
Butler, Clifford Charles (1922–1999), 15, 16, 28.

Cabibbo, Nicola (1935–2010), 25, 166.
Caesar, Gaius Julius (100 BC – 44 BC), 100.
Cantor, George Ferdinand Ludwig Philipp (1845–1918), 221.
Cara Romeo, Giovanni, 35.
Carnap, Rudolph (1891–1970), 212.
Charles VII (1403–1461), 100.
Charpak, George (1924–2010), 63, 114.
Chew, Geoffrey F., 131, 132.
Chilingarov, Artur Nickolayevich, 139.
Chu, Paul C.W., 139.
Cicero, Marcus Tullius (107 BC – 43 BC), 156.
Cifarelli, Luisa, 35, 56, 234.
Clark, Robert Alfred, 139.
Cohen, Paul (1934–2007), 83, 221.
Coleman, Sidney (1937–2007), 17, 139.
Contin, Andrea, 35.
Conversi, Marcello (1917–1988), 25, 41, 42, 147, 207.
Cooper, Leon N., 195, 196.
Cooper, W.A., 19.
Corbino, Orso Mario (1876–1937), 94.
Cosandey, Maurice Roger, 139.

D'Alì, Giacomo, 35.
Dadda, Luigi (1923–2012), 114.
Dausset, Jean-Baptiste-Gabriel-Joachim (1916–2009), 139, 140.
Deng Xiao Ping (1904–1997), 112, 137, 167.
Di Cesare, P., 35.
Dirac, Paul Adrien Maurice (1902–1984), 1, 47, 48, 49, 51, 95, 99, 100, 113, 114, 125, 126, 129, 137, 139, 164, 172, 178, 181, 188, 189, 208, 228, 230, 231, 235.
Doll, Richard (1912–2005), 139, 140.
Drell, Sidney D., 139, 140, 165.
Duby, Georges (1919–1996), 139, 140.
Duff, Michael James, 1, 7.
Dyson, Freeman, 128.

Eccles, John Carew (1903–1997), 114, 139.
Einstein, Albert (1879–1955), 3, 79, 80, 81, 100, 105, 126, 176, 177, 178, 180, 193, 194, 195, 196, 231.
Eisenhower, Dwight D. (1890-1969), 20.

Eliott, Harry (1920–2009), 16.
Englert, François, 98.
Epimenides of Knossos (c. 600 BC – c. 501 BC), 215.
Esaki, Leo, 139.
Esposito, B., 35.

Faddeev, Ludvig D., 162.
Farley, Francis James Macdonald, 63.
Ferdinand, Archduke of Austria Franz (1863–1914), 100.
Fermi, Enrico (1901–1954), 9, 12, 17, 43, 94, 100, 125, 126, 134, 137, 138, 171, 175, 182, 184, 193, 195, 196, 207, 225, 226, 228, 229.
Fermi, Laura (1907–1977), 113.
Feynman, Richard Phillips (1918–1988), 67, 68, 111, 129, 134, 143, 166, 179, 209.
Filthuth, H., 19.
Fiorentino, Eugenio, 74.
French, J. Bruce (1921–2002), 179.
Friedman, Jerome Isaac, 139.
Fritzsch, Harald, 45.

Galilei, Galileo (1564–1642), 3, 11, 79, 80, 84, 100, 106, 107, 125, 126, 154, 155, 156, 188, 189, 195, 221, 232.
Garibaldi, Giuseppe (1807–1882), 88.
Garwin, Richard L., 63, 114, 139, 140.
Gayther, D.B., 15.
Gatto, R., 166.
Gell-Mann, Murray, 15, 16, 17, 18, 29, 31, 44, 45, 131, 132, 139, 140.
Gianotti, Fabiola, 98.
Giusti, Paolo, 35.
Glashow, Sheldon Lee, 139, 140, 162.
Gödel, Kurt (1906–1978), 3, 73, 77, 78, 79, 81, 82, 83, 84, 174, 175, 177, 178, 186, 187, 188, 212, 213, 215, 216, 217, 218, 219, 221.
Goldbach, Christian (1690–1764), 82.
Goldberger, Marvin Leonard (1922–2014), 131, 132.
Gorbachev, Mikhail Sergeyevich, 100, 112, 137, 139, 228.
Gregory, Bernard Paul (1919–1977), 182.
Gribov, Vladimir Naumovich (1930–1997) 32, 35, 38, 127.
Gross, David Jonathan, 128, 133.
Guralnik, Gerald Stanford (1936–2014), 98.

Hagen, Carl Richard, 98.
Harbeke, Günther (1929–1989), 168.
Hauptman, Herbert Aaron (1917–2011), 139.
Heisenberg, Werner Karl (1901–1976), 131, 175, 176, 177, 178, 183, 219.
Hess, Victor Franz (1883–1964), 100, 124, 126.
Heuer, Rolf-Dieter, 98.
Higgs, Peter Ware, 98, 152, 162.
Hilbert, David (1862–1943), 174, 188, 213, 214, 215, 216, 217, 219, 237.
Hitler, Adolf (1889–1945), 81, 100, 178.
Hodgkin, Dorothy Mary Crowfoot (1910–1994), 26, 139.
Hubel, David H. (1926–2013), 139.
Huber, Robert, 139.

Ichimiya Tarao, 206.
Incandela, Joe, 98.
Izrael, Yuri Antonievich (1930–2014), 163.

James, G.D., 14, 15, 16.
Jentschke, Willibald Karl (1911–2002), 182.
John Paul II, Pope, Karol Józef Wojtyla, (1920–2005), 100, 139, 163.
Joliot, Frédéric (1900–1958), 184.
Joliot-Curie, Irène (1897–1956), 184.

Kabir, P.K., 166.
Kapitza, Pyotr Leonidovich (1894–1984), 1, 94, 95, 100, 113, 137, 139, 195, 228.
Karle, Jerome (1918–2013), 139, 140.
Kemeny, John George (1926–1992), 186, 217.
Kendall, Henry Way (1926–1999), 139.
Kibble, Sir Thomas Walter Bannerman, 98, 111.
Kobayashi, Makoto, 74, 75, 76.
Kohl, Helmut Josef Michael, 139.
Koshiba, Masatoshi, 139, 162.
Krool, Norman (1922–2004), 179.

Lamb, Willis Eugene (1913–2008), 178, 179, 209.
Landau, Lev Davidovich (1908–1968), 29, 128, 129, 131, 132.
Langevin, Paul (1872–1946), 94, 95.
Lattes, Cesare Mansueto Giulio (1924–2005), 43, 147, 207.
Lederman, Leon Max, 29, 62, 97, 136.
Lee, Tsung Dao, 25, 28, 29, 30, 31, 113, 139, 140, 167, 168, 196.
Lee, Yuan Tseh, 139.
Lehn, Jean-Marie, 139, 140.
Lenin, Vladimir Ilyich Ulyanov (1879–1924), 100.
Leutwyler, Heinrich, 45.
Lipatov, Lev Nikolaevich, 32, 33, 38.
Lipkin, Harry J. (1921–2015), 181.
Lipscomb Jr., William Nunn (1919–2011), 139.
Lobačskij, Nikolaj Ivanovich (1792–1856), 214.
Lorentz, Hendrik Antoon (1853–1928), 100, 126, 235.
Louis XVI (1754–1993), 100.

Mach, Ernst (1838–1916), 9, 78, 187, 189.
Magnéli, Arne (1914–1996), 139, 140.
Mainardi, Francesco, 74.
Majorana, Ettore (1906–1938 ca), 114, 188, 225, 226.
Mandelstam, Stanley, 131.
Mandula, Jeffrey Ellis, 17.
Marcellus, Marcus Claudius (c. 268 BC – 208 BC), 156.
Maskawa, Toshihide, 74, 75, 76.
Mason, Sir John (1923–2015), 16.
Massam, Thomas, 35, 74.
Mattarella, Piersanti (1935–1980), 164, 167, 169.
Mattarella, Rt. Hon. Bernardo (1905–1971), 165, 166, 167.

Mattarella, Sergio, 167, 168.
Matthews, Paul Taunton (1919–1987), 16.
Maxwell James Clerk (1831–1979), 81, 100, 126.
McGee, James Dwyer, (1903–1987), 16.
Miller, Arthur I., 168.
Monari, L., 74.
Montagnier, Luc Antoine, 139, 140.
Mössbauer, Rudolf Ludwig (1929–2011), 139, 164.
Muirhead, Hugh (1925–2007), 43, 147, 207.
Müller, Karl Alex, 139, 140, 164, 196.
Muller, T, 63.
Müller, Wilhelm Carl Gottlieb (1880–1968), 178.

Napoleon Bonaparte (1769–1821), 100.
Ne'eman, Yuval (1925–2006), 31, 114.
Neddermeyer, Seth Henry (1907–1988), 41, 147, 206.
Neurath, Otto (1882–1945), 212.
Newth, J.A., 19.
Newton, Sir Isaac (1643–1727), 100, 108, 126.
Nikitin, R., 62, 67, 68.
Nishina, Yoshio (1890–1951), 206.

Occhialini, Giuseppe Paolo Stanislao (1907–1993), 24, 25, 43, 44, 47, 48, 147, 171, 178, 181, 207, 208, 209, 210.
Oehme, Reinhard (1928–2010), 28, 29, 30, 31.
Okubo, Susumu, 44.
Oppenheimer, J. Robert (1904–1967), 9, 114, 188, 225, 226, 228.

Pais, Abraham (1918–2000), 29.
Palme, Olof (1927–1986), 137, 164, 169.
Palmonari, Federico, 35, 74.
Pancini, Ettore (1915–1981), 41, 42, 147, 207.
Paul, Wolfgang (1913–1993), 182, 183.
Pauli, Wolfgang (1900–1958), 12, 128, 182, 183.
Pauling, Linus Carl (1901–1994), 139, 140.
Pertini, Sandro (1896–1990), 137.
Petermann, André (1922–2011), 25, 56, 129, 139.
Petrucci, Guido, 19.
Piccioni, Oreste (1915–2002), 41, 42, 147, 207.
Piroué, Pierre A., 114.
Planck, Max Karl Ernst Ludwig (1858–1947), 100, 103, 107, 108, 126, 157, 177, 178, 184.
Poisson, Siméon Denis (1781–1840), 91.
Poma, Mario (1939–2015), 140.
Pomeranchuk, Isaak Yakovlevich (1913–1966), 129, 131.
Powell, Cecil Frank (1903–1969), 43, 147, 207.
Preiswerk, Peter (1907–1972), 182, 184.
Pupillo, Paolo, 146.
Puppi, Giampietro (1917–2006), 42.

Qian, Jiadong, 139, 140, 167, 168.

Rømer, Ole Christensen (1644–1710), 107.
Rabi, Isidor Isaac (1898–1988), 3, 13, 20, 21, 25, 26, 85, 109, 113, 171, 172, 173, 184.
Ramsey, Norman Foster (1915–2011), 21, 139, 140.
Reagan, Ronald (1911–2004), 112, 137, 228.
Retherford, Robert Curtis (1912–1981), 209.
Rieben, Henri (1921–2006), 139, 140.
Rochester, George Dixon (1908–2001), 28.
Roversi Monaco, Fabio Alberto, 146.
Russell, Bertrand Arthur William (1872–1970), 3, 8, 9, 10, 11, 77, 78, 79, 80, 81, 82, 84, 85, 185, 213, 214, 215, 216, 217, 218.
Rutherford, Lord Ernest (1871–1937), 24, 94, 95, 100, 126.

Sagdeev, Roald Zinnurovich, 139, 140.
Sakharov, Andrei Dmitrievich (1921–1989), 139.
Salam, Mohammad Abdus (1926–1996), 10, 16, 71, 139, 140, 167, 168.
Salmeron, Roberto A., 19.
Samuelsson, Bengt I., 139.
Sartorelli, Gabriella, 35.
Schopper, Herwig Franz, 167, 168.
Schrieffer, John Robert, 196.
Schrödinger, Erwin (1887–1961), 184.
Schwinger, Julian Seymour (1918–1994), 139, 140, 179, 209.
Sens, Johannes C. (1928–2008), 63.
Siegbahn, Kai Manne Börje (1918–2007), 139, 140, 167, 168.
Shore, Graham M., 162.
Sommerfeld, Arnold (1868–1951), 177, 178.
Spitaleri, Carmelo, 140.
Stalin, Joseph Vissarionovich (1879–1953), 1, 100.
Stevenson, Edward C., 206.
Street, Jabez Curry (1906–1989), 206.
Stueckelberg, Ernst Carl Gerlach (1905–1984), 129, 181.
Sun, Honglie, 139.
Szyszko, Jan, 139.

't Hooft, Gerardus 33, 38, 39, 40, 45, 139, 162, 236.
Takeuchi, Masa, 206.
Tarjanne, Pekka (1937-2010), 166.
Teller, Edward (1908–2003), 112, 114, 139, 207, 228.
Thomson, Sir Joseph John (1856–1940), 96, 180, 235.
Ting, Samuel C.C., 25, 139.
Trudeau, Pierre (1919–2000), 137.
Turing, Alan (1912–1954), 88, 92.

Valenti, Giuliano, 35.
van Rood, Jon J., 139, 140.
Velikhov, Eugenij, 112, 139, 140.
Veltman, Martinus J.G., 236.
Veneziano, Gabriele, 36, 38, 39.

259

Veronesi, Umberto, 139, 140.
Villi, Claudio (1922–1996), 25, 139, 140.
von Hindenburg, Paul (1847–1934), 100.
von Neumann, John (1903–1957), 81, 83, 173, 174, 175, 188, 218.

Weisskopf, Victor Frederick (1908–2002), 25, 32, 34, 38, 48, 52, 53, 55, 60, 68, 109, 139, 165, 166, 178, 179, 182, 183, 192, 207, 208, 209, 225, 231.
Weyl, Claus Hugo Herman (1885–1955), 49, 208.
Wheeler, John Archibald (1911–2008), 78, 131, 184, 185, 186.
Whitehead, Alfred North (1861–1947), 81, 82, 213, 214, 216, 217, 219.
Wick, Gian Carlo (1909–1992), 115.
Wightman, Arthur Strong (1922–2013), 168.
Wigner, Eugene Paul (1902–1995), 9, 81, 83, 113, 114, 129, 173, 174, 175, 188, 198, 218, 219, 225, 226, 228, 231.
Wiik, Bjorn H. (1937–1999), 25, 162.
Wilson, (Bob) Robert R. (1914–2000), 231.
Wilson, Charles Thomson Rees (1869–1959), 94, 95, 176.
Wilson, Richard, 139.
Witten, Edward, 162.
Wolf, Gunter, 25.
Wright, Sir Charles Seymor (1887–1975), 91.
Wright, William David (1906–1997), 16.
Wu, Chien Shiung (1912–1997), 25, 29, 84, 85, 139, 140, 172, 183.
Wu, Maw-Kuen, 139.

Yang, Chen-Ning, 28, 29, 30, 31.
Yeltsin, Boris Nikolayevich (1931–2007), 100.
Yonath, Ada E., 139.
Yukawa, Hideki (1907–1981), 42, 76, 100, 130, 206, 207.

Zel'dovich, Yakov Borisovich (1914–1987), 61.
Zhou Guang Zhao, 112.
Zweig, George, 166.

ANALYTIC INDEX OF THE MAIN TOPICS

ABJ anomaly, 45.
AdA, 135, 136.
ADONE, 23, 72, 76, 135, 136.
AFB phenomena, 143, 144, 145, 147.
Algebra, 215, 238.
ALICE, 116.
Analysis, 215, 238.
Annihilation, 52, 75, 76, 208, 231, 236.
   Annihilation phenomena, 188, 209, 231, 236.
   Annihilation, (e^+e^-), 35, 36, 37, 49, 51, 208.
   Annihilation, ($\bar{p}p$), 23, 30, 37, 72, 74, 75.
   Annihilation processes, 47, 50, 187, 208, 209.
   Annihilation virtual, 188, 209, 231, 236.
Antimatter, 108, 127, 128, 175, 183, 188, 189, 235.
Antiparticle, 14, 18, 48, 49, 76, 209, 230, 231, 235, 236.
Arithmetic, 82, 215, 216, 217, 238.
Asymptotic freedom, 27, 28, 39, 99, 133, 134, 162, 187, 200.
Atomic, 9.
   Atomic cloud, 48.
   Atomic matter, 12.
   Atomic nucleus, 95, 225.
   Atomic volume, 95.
Axiom, 82, 174, 218, 219.
   Axiom, logic, 177.
   Axiom, system, 82, 174, 215, 219.

Baryon Λ^0, 15, 18, 28, 31, 32, 36, 130.
Baryonic number, 181.
Baryonic state, 31, 95.
Beam, 30, 73, 192.
   Beam charged, 128.
   Beam neutral, 30.

Berlin Wall, 112.
Big Bang, 56, 60, 102, 103, 108, 185, 230.
   Big Bang, ex-, 56.
   Big Bangs, two, 102, 103, 106, 157.
   Big Bangs, three, 3, 93, 100, 102, 103, 104, 106, 108, 157.
Blackett, effect, 3, 5, 68, 87, 88, 89, 90, 92.
   Blackett group, 5, 7, 8, 9, 10, 12, 14, 18, 42, 54, 61, 65, 68, 79, 85, 93, 148, 156, 172, 191, 231.
   Blackett Institute, 4, 6, 110, 111, 140, 164, 167, 169.
   Blackett scholarship and diplomas, 158.
Black Holes, 60, 78, 219.
Bomb, atomic, 9.
   Bomb, nuclear, 227.
   Bomb, nuclear fission, 9.
   Bomb, nuclear fission, plutonium, 11.
   Bomb, nuclear fission, uranium, 11, 95.
   Bomb, nuclear fusion, H-bomb, 11, 95, 100, 202, 229.
Boson, 17, 98, 104, 105, 187, 193, 194, 195, 197, 198, 208, 236.
Bosonic dimension, 105, 106.

Cambridge Circle, 3, 93, 95, 212, 213.
   Blackett Circle, 95.
   Dirac Circle, 95.
   Kapitza Circle, 95.
CERN, 3, 5, 6, 8, 10, 13, 19, 20, 21, 23, 25, 26, 30, 34, 35, 36, 39, 44, 61, 65, 67, 69, 70, 72, 73, 74, 75, 84, 85, 97, 98, 116, 123, 147, 154, 167, 171, 172, 175, 181, 182, 183, 184, 189, 191, 192, 225, 230, 234.
Cervinia Lab (3,480 meters a.s.l.), 13, 20.
Charge, 29, 52, 129, 171, 180, 194, 203, 206.
   Charge, colour-electric QCD, 40.
   Charge, colour-magnetic QCD, 39, 40.
   Charge, conjugation, 27, 172.
   Charge, electric, 44, 97, 171, 180, 181, 201.
   Charge, flavour, 104, 223.
   Charge, gauge, 104, 223.
   Charge, nuclear, 201.
   Charge, strange, strangeness, 3, 19, 171, 175, 185, 191.
   Charge, strong, 69, 76, 203.
   Charge, weak, 6, 69, 203.
Collider, 23, 34, 35, 36, 39, 72, 76, 135, 154, 181, 192, 230.
Collision, 34, 35, 36, 127, 134, 192, 201.
Complex nature of space-time, 235.
Complex systems, 10, 141, 142, 145, 156.
Complexity, 99, 141, 143, 145, 147, 156.
   Complexity, asymptotic limit of, 99.
   Complexity, highest limit, 99, 101, 145, 156.
   Complexity, lowest limit, 99, 101, 145, 156.
Confinement, 27, 28, 39, 40, 187, 200.
Cosmic rays, 3, 13, 95, 99, 100, 117, 123, 124, 126, 148, 171, 175, 176, 178, 187, 191, 206, 207.

Cosmic rays "meson", 41, 42.
CP and T invariances, 29, 30, 31, 83, 173, 175, 183.
    CP breaking, 27, 28, 29, 31, 127.
    CP violation, 27, 29, 127, 231.

Decay mode, 29, 30, 44.
Deep Inelastic Scattering (DIS), 37, 38, 55.
Dirac, equation, 23, 47, 100, 126, 179, 181, 208, 210, 230, 231, 235.
    Dirac, Lecture Hall, 164, 169.
    Dirac, mathematical formalism, 49.

EBUS (Evolution of our Basic Understanding of the laws governing the world in its Structure), 99, 100.
EEE, 123, 124.
Effective Energy, 27, 28, 32, 33, 34, 35, 36, 37, 38, 39, 127, 134, 187, 203, 204.
EGM (Evolution of Gaugino Masses) effect, 55, 57, 59, 60, 127.
Einstein equation, 180.
Electricity, 96, 100, 126, 148.
Electromagnetism, 61, 81, 133.
Electron, 12, 48, 49, 50, 52, 53, 61, 62, 63, 64, 66, 72, 73, 74, 76, 96, 97, 172, 176, 179, 180, 181, 188, 195, 196, 197, 198, 201, 206, 209, 222, 229, 230, 232, 235.
    Electron masses, 61, 206.
    Electron pair (Cooper-pair), 195, 196.
    Electron-antielectron annihilation, 76.
    Electron-positron pair, 208, 210, 236.
    Electron spin, 195.
ELN (Euroasiatic Long Intersecting Storage Accelerator), 56, 112, 146, 152, 153, 154.
EMFCSC, 5, 6, 8, 10, 11, 12, 26, 109, 114, 138, 160, 169.
    EMFCSC, activities and schools, 6, 68, 114, 159, 165, 167, 225.
    EMFCSC-Ettore Majorana prize, 138.
Energy, 20, 104, 107, 134, 135, 136, 154, 177, 180, 191, 192, 200, 201, 202, 223, 225.
    Energy density, 96.
    Energy, E_{GUT}, 54, 58, 60.
    Energy, E_{SU}, 54.
    Energy high physics, 13, 35, 39, 47, 95, 117, 129, 132, 171.
    Energy level, 20, 54, 60.
    Energy level, hydrogen, 52, 53, 60, 209.
    Energy, mass, 147, 202.
    Energy nominal, 34, 35, 36, 37, 192, 203, 204.
    Energy, Planck, 54, 57, 58, 60.
    Energy, rest-mass, 208.
    Energy, vacuum, 180.
EPS, 39, 135, 136.
Equation, 57, 180.
Erice Statement, 95, 115, 137, 138, 139, 228.
η–η' mixing, 45.
η–η' problem, 36, 45.
EWRL (Evolution of the World in its Real Life), 99, 100.
Experimental proof, 5, 12, 18, 19, 29, 69, 83, 84, 97, 106, 128, 133, 134, 143, 145, 156, 171, 175, 178, 183, 189, 195, 198, 206, 207, 208, 209, 212, 213, 220.

Family, first, 43, 230.
    Family, second, 28, 43.
    Family, third, of fundamental particles, 24, 77.
    Family, third, three, 3, 24, 25, 41, 42, 43, 73, 75, 76, 77.
Fermi, couplings, 6, 42.
    Fermi energy, 104, 190.
    Fermi force, 43, 94, 190.
    Fermi group, 184.
Fermions, 17, 31, 36, 104, 105, 133, 187, 193, 194, 194, 195, 197, 198.
Flat magnet, 62, 65, 67, 68.
Flavour mixing, 27, 28, 30, 31, 151.
    Flavour mixing in the leptonic sector, 224.
    Flavour mixing in the quark sector, 224.
Forces, electromagnetic, 10, 12, 81, 105, 133, 143, 173, 175, 190, 208.
    Forces, electroweak, weak, 12, 94, 100, 104, 105, 126, 130, 133, 148, 175, 190, 201, 203.
    Forces, fundamental, 5, 10, 12, 57, 84, 104, 144, 148, 149, 187, 190, 201, 208, 209, 223, 236.
    Forces, gravitational, 12, 54, 60, 105, 149, 175, 177, 190.
    Forces, non abelian gauge e non, 133, 151, 162.
    Forces, nuclear, 12, 41, 43, 44, 128, 130, 133, 143, 202, 206, 207, 226.
    Forces, strong, subnuclear, 38, 104, 128, 130, 175, 190, 199, 208.
    Forces, unification of the fundamental forces of Nature, 3, 5, 47.
Form factors, 131.
Function, 131, 184, 238.
    Function, β, 162.

Gamma-ray, γ–ray, photon, 12, 48, 51, 208, 209.
GAP, 3, 47, 54, 55, 58, 60, 127, 224.
Gauge, boson, 208, 231, 236.
    Gauge couplings, 54, 55, 56, 57, 231.
    Gauge forces, renormalization, 231, 236.
    Gauge interaction, 32.
    Gauge principle, 151.
    Gauge unification, 54, 55, 60.
Geometry, 215, 238.
    Geometry, elliptical 214.
    Geometry, Euclidean, 176, 214, 215.
    Geometry, hyperbolic, 214.
    Geometry, non-Euclidean, 215.
Glue, 104, 132, 193, 194, 197, 198.
    Glue, nuclear, 3, 12, 22, 41, 42, 43, 44, 46, 76, 100, 187, 206, 207.
Gluon, 32, 36, 45, 46, 132, 134, 236.
    Gluon jets, 3, 41, 43, 46.
    Gluon quantum numbers, 46.
    Gluon-induced jets, 46.
Gran Sasso Underground Laboratory, 54, 58, 117, 147, 154, 201.

HERA, 154.
Hierarchy, problem, 208, 224.

High and Low transverse momentum, P_T, 34, 37, 38.
Hiroshima and Nagasaki, 9, 11, 95, 225.
    Hiroshima, Cultural, 229.
    Hiroshima, Political, 229.
Holism, 155.

ICSC-World Lab, 8, 113, 115, 116.
ICTP, 10, 11, 12.
Imagination, 26, 107, 173, 174, 188, 196, 232, 233.
Immanent, 83, 188, 220.
Imperial College, 1, 2, 7, 16, 90, 111, 172, 184.
Indecidability, 219.
Infinity, 83, 188, 221.
INFN, 13, 74, 135.
Instantons, 3, 41, 43, 45, 151, 224.
Interactions, 32, 33, 34, 35, 37, 38, 61, 62, 107, 127, 130, 173, 200, 203, 204.
    Interactions electromagnetic, 32, 133, 173, 198.
    Interactions gauge, 32.
    Interactions hadronic, 36, 40.
    Interactions nuclear, 41, 130.
    Interactions particle, 10, 20, 173.
    Interactions proton, 34.
    Interactions strong, 32, 37, 42, 76, 128, 129, 131, 132, 133, 172.
    Interactions weak, 29, 32, 130, 133, 187, 190, 201.
International Seminars on Nuclear War, 115, 119, 123.
Isobosons, 17.
Isofermions, 17.
Isosinglet, 45.
ISR (Intersecting Storage Rings), 13, 34, 36, 39, 134, 191, 192.
    ISR, nominal energy, 34, 35, 37.

Jungfraujoch Lab (3,580 meters a.s.l.), 7, 8, 18, 20, 80, 172.

LAA (Large Asymmetry Analyser), 116, 154.
Lagrangian, 96.
Lamb-shift, 22, 47, 53, 81, 178, 179, 209.
Language, 9, 83, 102, 103, 106, 156, 157, 212, 213, 214, 215, 216, 217, 222.
    Language mathematical, 217.
    Language written, Permanent Collective Memory (PCM), 102, 106, 107, 155, 156.
Law, 11, 141, 202.
    Law, acoustic, 143, 144.
    Law, bosonic, 104.
    Law, conservation, 181.
    Law, energy conservation, 12.
    Law, fermionic, 104.
    Law, fundamental invariance, 198.
    Law, fundamental of Nature, 11, 141, 143, 173, 185.
    Law, physics, 83, 173, 185, 203.
    Law, statistical, 104, 193.
        Law, Bose-Einstein statistic, 126, 193, 194, 195, 196.

Law, Fermi-Dirac statistic, 100, 126, 193, 195, 196.
LEP (Large Electron Positron Collider), 13, 154, 181, 188, 191, 192, 223, 230, 234.
LEP white book, 154.
Lepton, physics, 41, 104, 181, 193, 194, 195, 197, 198.
    Lepton, heavy, 25, 42, 43, 48, 74, 75, 76, 77.
    Lepton, third, (HL), 3, 23, 25, 30, 42, 44, 71, 73, 74, 75, 76, 77, 78, 127, 135.
LHC (Large Hadron Collider), 13, 60, 154, 181, 188, 191, 192, 234.
Light, 190, 198, 202.
    Light, speed, 106, 107, 108.
Logic (rigorous theoretical logic) mathematic, 77, 78, 82, 83, 102, 106, 107, 155, 156, 157, 175, 186, 188, 213, 214, 215, 216, 219, 223, 238.
    Logic of Nature, 9, 12, 79, 80, 83, 98, 99, 141, 156, 175, 233.
    Logic, 1, 102, 103, 106, 157.
Logical variables, 216.

Machine at CERN, 175, 191, 192, 230.
Magnet, 6-metre, 66, 67.
Magnetism, 100, 126, 148.
Majorana neutrinos, 225.
Manhattan Project, new, 9, 10, 11, 123, 181, 188, 225, 227, 228, 229.
Mass, 44, 45, 61, 77, 78, 95, 104, 107, 147, 162, 180, 181, 194, 202, 206, 223.
    Mass, imaginary, 39, 96, 98, 224.
    Mass, η', 45.
    Mass, η^0, 44.
    Mass, K, 45.
    Mass, missing, 224.
    Mass, third lepton, 75, 76.
    Mass, pion, 45.
    Mass, X^0-, 44.
Matter, 12, 41, 107, 143, 180, 181, 189.
    Matter, inert, 101, 102, 106, 188, 220, 223.
    Matter, living, 101, 102, 106, 188, 219, 220, 223.
    Matter, nuclear, 207.
    Matter, stability, 54, 180, 223.
    Matter-antimatter, symmetry, 129, 224.
Maxwell equations, 81.
Mendeleev table, 48, 94.
Meson, 17, 29, 31, 32, 36, 42, 44, 45, 76, 130, 206.
    Meson θ^0, heavy, 6, 14, 16, 17, 18, 19, 21, 28, 42, 126, 171, 191.
    Meson long lived, 30.
    Meson mixing, 127, 134.
    Meson, pseudoscalar, 44, 45, 134.
    Meson, K-, 83, 231.
    Meson, K^0-, 44, 45, 175.
    Meson, μ, 207.
    Meson, pi-, (π-meson), 3, 12, 15, 22, 41, 42, 43, 44.
    Meson, S^0-, 97.
    Meson, scalar, 97.
    Meson, strange, 175.
    Meson, X^0-, 45.

Meson, η−, 44, 45.
Meson, η'−, 45, 46.
Meson, vector, 134.
Mesonic state, 31, 32, 95.
Minimal life, 101, 102, 223.
Mixing angle, (η−η'), 44, 45.
Motor of progress, 9, 109, 112.
Multihadronic final state, 33, 34, 38.
Muon, 42, 48, 61, 62, 64, 66, 67, 71, 72, 73, 74.
    Muon, heavy electron, 3, 42, 60, 61.
    Muon, electromagnetic properties, 64, 65, 67.
    Muon lifetime, 69.
    Muon, anomalous magnetic moment of the, $(g-2)_\mu$, 21, 48, 60, 61, 62, 63, 64, 67, 179, 189.

NBC technology, 97.
Neutrino, 12, 23, 25, 74, 75, 201, 202, 225, 231.
    Neutrino mixing, 54.
Neutron, 12, 15, 75, 76, 180, 181, 184, 201, 202, 206.
Newton constant, 108.
Non invariance of symmetry operators, 29.
NP-complete, 155, 156.
Nuclear fission, 9, 11, 95, 225.
Nuclear fusion, 1, 11, 95, 202.
Nucleon, heavy, 76, 206.
Nucleus, 48, 94, 100, 126, 207.
    Nucleus atomic, 95.
    Nucleus helium, 202.
Numerical variables, 216, 217, 238.

Optics, 100, 126, 148.

Pair production, (e^+e^-), 7, 18, 19, 21, 47, 48, 51, 74, 99, 171, 175, 191, 208, 210.
Panisperna group, 94.
PAPLEP (Proton AntiProton Annihilation into LEpton Pairs) experiment, 30.
Paradox, twin, 95.
    Paradox, assertions, 215.
Paramagnetism, 95.
Particle, 10, 12, 14, 17, 18, 20, 24, 27, 32, 34, 36, 39, 42, 48, 49, 52, 74, 77, 104, 105, 175, 176, 179, 190, 197, 201, 203, 207, 208, 222, 223, 230, 231, 236.
    Particle, elementary, 10, 32, 75, 76, 83, 84, 96, 105, 130, 166, 173, 198, 198, 204, 235.
    Particle, charge, 134, 180, 204.
    Particle, God, 96, 97, 98.
    Particle, physics, 17, 128, 133.
    Particle, QED-, 71.
    Particle, spin, 194.
    Particle, strange, 5, 14, 16, 28, 69, 81, 95, 100, 125, 126, 148, 172.
    Particle, supersymmetric, 60.
    Particle, useless, 150, 188, 230.

Particles, V, 3, 7, 12, 14, 15, 22, 27, 28, 31, 171.
Photino, 198.
Photographic emulsion, 147, 207.
Photon, antiphoton, 12, 49, 50, 53, 63, 64, 198, 208, 209, 236.
Physics, 1, 5, 6, 20, 21, 27, 73, 77, 131, 147, 171, 173, 178, 189, 203, 212, 219, 224.
    Physics, atomic, 9, 78, 187, 189.
    Physics, Black Holes, 78.
    Physics, deconfined colour charge, 224.
    Physics, discovery, 6, 68, 94, 125, 137, 186.
    Physics, experimental, 8, 112, 124, 129.
    Physics, frontiers, 9, 17, 30, 60, 80, 85, 169, 176.
    Physics, future, modern, 53, 68, 94, 235.
    Physics, high energy, 47, 117, 132, 184.
    Physics, high precision, 20, 21, 61.
    Physics, imaginary masses, SSB, 151, 152, 224.
    Physics, K decay, 29.
    Physics, lepton, 41.
    Physics, meson, 44, 231.
    Physics, nuclear, 13, 144, 148.
    Physics, Planck scale, 224.
    Physics, quantities, 177.
    Physics, strange particles, 14, 16.
    Physics, subnuclear, 13, 146, 154, 225.
    Physics, theoretical, 111, 112, 128, 129, 178.
    Physics, virtual, 3, 5, 22, 32, 47, 54, 55, 60, 61, 63, 64, 98, 99, 187, 208.
    Physics, virtual phenomena, 98, 225.
Planck, action, 107, 108, 175.
    Planck constant, 78, 177, 194.
    Planck energy, E_{Planck}, 54, 58, 60.
    Planck length, 108.
    Planck scale, 57, 224.
    Planck time, 108.
Planetary Emergencies, 114, 115, 116, 122, 123, 228, 229.
Poisson distribution, 91.
Polynomial magnetic fields, 62, 67.
Principia Mathematica, 81, 82, 213, 215, 216, 217, 219.
Principle, 23, 25, 44.
    Principle Invariance, 173.
    Principle Relativity, 79, 80.
    Principle the third excluded, 82, 177, 178, 219.
    Principle uncertainty, 175, 176, 177, 178, 219.
    Principle verifiability, 212.
Proton, 12, 34, 48, 73, 74, 75, 76, 131, 134, 180, 181, 190, 191, 192, 201, 202, 203, 204, 206.
PS (Proton Synchrotron), 13, 30, 191, 192.
Pseudoscalar ($q\bar{q}$) system, 45.
Pseudoscalar meson, 44, 134.
Pseudoscalar meson $SU(3)_f$ multiplet, 45.
Pseudoscalar η, 44, 45.
Puppi-"triangle", 42.

QCD (Quantum ChromoDynamics), 3, 27, 32, 33, 36, 38, 39, 43, 45, 134, 143, 144, 149, 162, 200, 236.
   QCD, confinement, 127.
   QCD, light, 127.
   QCD non-perturbative effect, 34, 46, 224.
   QCD, hidden side, 3, 27, 32, 40, 134.
QED (Quantum ElectroDynamics), 43, 61, 63, 64, 71, 83, 129, 144, 149, 172, 175, 179, 208, 236.
QFD (Quantum Flavour Dynamics), 149, 236.
Quantum Mechanics, 175.
Quantum number, 14, 15, 16, 17, 18, 31, 46, 73, 95, 97.
Quark, 32, 36, 39, 40, 43, 44, 45, 73, 75, 104, 132, 134, 181, 193, 194, 195, 197, 198, 200, 203, 236.
   Quark confinement, 39.
   Quark family, 28, 76.
   Quark flavours, 31.

Radiative decays, (J/ψ), 45.
Radiative effect, 55.
Radioactive decay processes, 12.
Radioactivity, 94, 95, 100, 148, 178.
   Radioactivity, artificial, 184.
Reason, 102, 106, 107, 219, 223.
Reductionism, 155.
Renormalization, 129, 162, 236.
   Renormalization group equations (RGEs), 152.

SC, 13, 175, 191, 192.
Scattering amplitudes, 131.
Scaling, 39, 134.
Science (Rigorous experimental logic), 1, 23, 77, 78, 83, 89, 90, 92, 99, 100, 102, 103, 104, 106, 107, 108, 109, 113, 123, 124, 125, 134, 145, 156, 157, 186, 217, 219, 223.
   Science, creativity, 83.
   Science, ethics, 1, 3, 23, 25, 71.
   Science, for peace, 112, 137, 138, 140, 227, 228.
   Science, frontier, 10, 117.
   Science, Galilean, first level, 9, 84, 102, 107, 108.
   Science, role, 8, 9, 109.
Scientific Institutions, 3, 5, 8, 10, 20, 21, 69, 84, 109.
Seven components, 103, 104, 157, 223.
Shimming technology, 65, 67, 68.
SLAC (Stanford Linear Accelerator Center), 39, 76, 134.
Slow neutron technology, 94, 184.
Space, 97, 104, 105, 106, 223, 235.
   Space, Intrinsic- (isospin), 17, 96, 97.
   Space-like, 100, 126.
   Space-Time, 103, 147, 198.
   Space-Time, dimensions, number of expanded, 105, 106, 143, 144, 157, 198, 216.
   Space-Time, properties, 126, 198.
   Space-Time, Lorentz-, 17, 96, 97, 235.

SPEAR (Stanford Positron Electron Accelerating Ring), 37, 136.
Sphinx Observatory, 7, 8.
Spin, 44, 97, 104, 194, 195, 197, 198, 223.
SPS (Super Proton Synchrotron), 13, 191.
Standard Model & Beyond (SM&B), 148, 149, 150, 151, 152, 153, 231.
    Standard Model, 126, 149, 151, 199, 230, 231, 233, 236.
Statements,
    Bergmann, Peter Gabriel, 180.
    Blackett, Patrick M.S., 40, 81, 96, 111.
    Chew, Geoffrey F., 132.
    Cicero, Marcus Tullius, 156.
    Corbino, Orso Mario, 94.
    Einstein, Albert, 105, 176.
    Fermi, Enrico, 12, 134, 138, 182, 225, 226.
    Feynman, Richard P., 0, 68, 111, 129.
    Galilei, Galileo, 80, 154, 195, 232.
    Gell-Mann, Murray, 132.
    Goldberger, Marvin Leonard, 132.
    Gribov, Vladimir N., 32, 35, 36.
    Gross, David J., 128, 132, 133.
    Hilbert, David, 215.
    Kemeny, John G., 217.
    Landau, Lev D., 128, 129.
    Lipatov, Lev N., 33.
    Pauli, Wolfgang, 182.
    Rabi, Isidor I., 26, 173.
    Russell, Bertrand A.W., 79, 80, 81.
    Sommerfeld, Arnold, 177, 178.
    't Hooft, Gerardus, 33, 39.
    Veneziano, Gabriele, 36, 39.
    von Neumann, John, 173, 174.
    Weisskopf, Victor F., 32, 33, 34.
    Wheeler, John Archibald, 185.
    Wigner, Eugene P., 173, 174, 218, 226.
    Wu, Chien Shiung, 25, 85, 172.
Strangeness, 3, 6, 14, 17, 18, 19, 21, 27, 69, 95, 171.
    Strangeness mixing, 29.
    Strangeness, quantum number, 15, 16, 17, 18.
Subnuclear Physics School, 68, 156, 165, 166, 169, 174, 181, 183.
Subnuclear universe, 3, 5, 9, 14, 27, 76, 143, 183, 190, 233.
Sun, 12, 190, 201, 202.
Superconductivity, 195, 196.
    Superconductivity cold, 196.
    Superconductivity warm, 196.
Superfluidity, 95, 187, 195.
Supergravity, 17.
Superspace, 103, 105, 106, 144, 157, 199.
Supersymmetry (SUSY), 17, 57, 59, 60, 105, 151, 187, 197, 224.
Superworld, 17, 54, 57, 59, 60, 103, 105, 143, 144, 146, 157, 187, 197, 199.
Symmetry, 105, 132, 166, 198, 224, 231.

Symmetry breaking, 236.
Symmetry operators C and P, breaking ($\neq$), 29, 127, 128, 129.
Symmetry operators, collapse, 129.
Symmetry $SU(3)_f$, global, 31.

Technology of high precision magnetic fields, 3, 6, 20, 62, 65, 67, 182.
(θ–τ) puzzle, 27, 28, 29, 129.
Theorem, 82, 173, 174, 175, 177, 219.
    Theorem, Gödel, 81, 82, 174, 175, 177, 213, 219.
    Theorem, CPT, 128, 129, 133.
    Theorem, No-go, 17.
    Theorem, Wigner, 10, 83, 105, 173, 198.
Theory, 45, 68, 111, 128, 129, 238.
    Theory, abelian gauge, 133.
    Theory, atomic, 189.
    Theory, Bohr, 176.
    Theory, CPT, 129.
    Theory, Grand Unified (GUT), 5, 47, 151, 152, 236.
    Theory, Lorentz-invariant, 129, 132.
    Theory, M-, 17.
    Theory, probability, 88, 89, 91.
    Theory, proximity, 238.
    Theory, Relativistic Quantum Field (RQFT), 127, 128, 129, 130, 131, 132, 133, 134.
    Theory, Relativistic Quantum String (RQST), 54, 151, 152.
    Theory, S-matrix, 127, 128, 129, 130, 131, 132, 133, 134.
    Theory, Yukawa field, 130, 207.
Three columns, 149, 230.
Time, 104, 105, 173, 175, 177, 198, 223, 235.
    Time, axis, 83, 48, 49, 50, 51, 52.
    Time-like, 75, 76, 100, 126, 131.
    Time-reversal invariance, 9, 10, 170, 173, 175, 198, 199.
    Time-reversal operator, 83, 173, 175.
TOF (Time Of Flight), 73, 128.
Topology, 215, 238.
Tower of Thought, Piersanti Mattarella, 3, 159, 160, 163, 167, 169.
Transcendent, 83, 185, 188, 220.
Transverse momentum, 34, 37, 38.

UEEC (Unexpected Events with Enormous Consequences), 96, 99, 100, 125, 126, 137, 143, 145, 147, 148.
    Unexpected discovery, 7, 12, 14, 46, 78, 96, 148.
    Unexpected phenomena, 177, 233.
Universality Features, 27, 28, 33, 34.
Universe, 60, 77, 78, 84, 102, 103, 157, 180, 181, 185, 190, 230.
    Universe, action of the, 108.
    Universe, age of the, 108.
    Universe, missing mass in the, 224.
    Universe, origin of the, 150, 188, 230.
    Universe, radius of the, 108.

Universe, vacuum of the, 102.

Vacuum polarization, 3, 22, 47, 52, 53, 126, 208.
　Vacuum polarization effects, 5, 47, 48, 52, 55, 60, 84, 99, 179, 208.
　Vacuum polarization in hydrogen, 48, 53, 178.
Vienna Circle, 78, 81, 84, 175, 177, 186, 187, 212, 215.
Violation ($\neq$) of Parity (P) and of charge conjugation (C), 27, 29, 172, 231.
Virtual History, 3, 93, 98, 99.
Virtual phenomena, 5, 47, 48, 98, 188, 189, 209, 210, 231, 235, 236.
Virtual process, 131, 209.

Weak processes, 29, 33, 42.
WFS (World Federation of Scientists), 8, 113, 114, 115, 120, 123, 138.
　WSP, PMP (Permanent Monitoring Panel), 120, 121, 122, 123.
Wilson cloud-chamber, 7, 13, 14, 20, 23, 95, 97, 147, 176.
World War, Second, 3, 5, 7, 9, 84, 87, 89, 92.

Yukawa's meson, 42, 76, 206, 207.
　Yukawa's particle, 207.

WE ALL SHOULD **PAY TRIBUTE** TO A PHYSICIST WHO HAS PLAYED A VITAL ROLE IN THE DISCOVERY OF THE SUBNUCLEAR UNIVERSE AND IN THE PROMOTION OF SCIENTIFIC CULTURE.

THE REAL MOTOR FOR THE PROGRESS IN TECHNOLOGICAL INVENTIONS THAT ALLOW THE QUALITY OF LIFE TO BE AT THE LEVEL IT IS TODAY IS THE SCIENTIFIC DISCOVERY AT THE 1st LEVEL OF GALILEAN SCIENCE.